THE DESIGN OF FRONTIER SPACES

The Design of Frontier Spaces

Control and Ambiguity

Edited by

Carolyn Loeb
Michigan State University, USA

Andreas Luescher
Bowling Green State University, USA

Routledge
Taylor & Francis Group
LONDON AND NEW YORK

First published 2015 by Ashgate Publishing

Published 2016 by Routledge
2 Park Square, Milton Park, Abingdon, Oxon OX14 4RN
605 Third Avenue, New York, NY 10017

First issued in paperback 2021

Routledge is an imprint of the Taylor & Francis Group, an informa business

British Library Cataloguing in Publication Data
A catalogue record for this book is available from the British Library.

Library of Congress Cataloging-in-Publication Data
The design of frontier spaces : control and ambiguity / [edited] By Carolyn Loeb and Andreas Luescher.
pages cm. -- (Design and the built environment)
Includes bibliographical references and index.
ISBN 978-1-4724-1976-7 (hardback) -- ISBN 978-1-4724-1977-4 (ebook) -- ISBN 978-1-4724-1978-1 (epub) 1. Place (Philosophy) in architecture. 2. Space (Architecture)--Psychological aspects. 3. Architecture--Psychological aspects. 4. Architecture and society. I. Loeb, Carolyn S., 1948- editor. II. Luescher, Andreas, editor.
NA2500.D475 2015
720--dc23

2014041054

ISBN 13: 978-0-367-78734-9 (pbk)
ISBN 13: 978-1-4724-1976-7 (hbk)

Contents

PART II BORDER BUILDINGS

PART III SPATIAL AMBIGUITY AND (DIS)EMBODIED MEMORY

List of Illustrations

Notes on Contributors

Eric Aronoff (PhD Rutgers University) is Associate Professor in the Residential College in the Arts and Humanities at Michigan State University, where his teaching and research interests include nineteenth- and twentieth-century American literature; anthropology, literature and literary criticism; science fiction; and literature and the environment. His book, *Composing Cultures: Modernism, American Literary Studies, and the Problem of Culture* (University of Virginia Press, 2013), examines the intersection of theories of culture in anthropology, literature and literary criticism in American modernism. Since earning his BA in comparative literature from Princeton University (1990) with concentrations in Hebrew and German literature, Aronoff has also had an intense interest in Israeli culture and society; he regularly visits and leads study abroad programs to the region.

Yael Aronoff is the Michael and Elaine Serling Chair of Israel Studies and Associate Professor of International Relations at James Madison College at Michigan State University. She received her PhD in political science from Columbia University (2001). Her book, *The Political Psychology of Israeli Prime Ministers: When Hard-Liners Opt for Peace*, was published by Cambridge University Press in 2014. Aronoff has published in *Political Science Quarterly*, *Israel Studies*, and *Israel Studies Forum* (now *Israel Studies Review*). Recent book chapters include "Predicting Peace: The Contingent Nature of Leadership and Domestic Politics in Israel," in *Democracy and Conflict Resolution: The Dilemmas of Israel's Peacemaking* (Syracuse University Press, 2014), edited by Miriam Fendius Elman, et al.

Giovanna Guidicini has studied architecture and conservation at Università di Ferrara. She obtained her PhD at the University of Edinburgh in architectural history in 2009, and was Teaching Fellow at the Edinburgh College of Art and at the University of Edinburgh between 2010 and 2012; since 2012 she is Lecturer in Architecture at Plymouth University. Her area of interest is the representation of power through triumphal entries and palatial celebrations in Europe during the early modern period, and how these concepts translate into temporary and permanent architecture. Guidicini has published book chapters and articles on this topic, and is currently working on her first monograph comparing the use of space during urban and palatial celebrations in Scotland.

Richard Kurdiovsky studied art history in Vienna, writing his thesis on the life and work of the architect Alfred Castellitz (2000) and his dissertation on Gottfried Semper's stylistic influence on the *oeuvre* of Carl von Hasenauer (2008). From 1997 to 2004 he was a freelance member of the staff at the Architecture Collection of the Albertina, Vienna. Since 2005 he has been on the staff of the Commission for the History of Art (now the Division for History of Art of the Institute for History of Art and Musicology) of the Austrian Academy of Sciences. Kurdiovsky is Assistant Lecturer at the University of Vienna and the Technical University of Vienna. Special areas of research include Vienna's Hofburg in the nineteenth century, Central European architecture and garden art from the Baroque to the early twentieth century, and the urban history and culture of the Habsburg monarchy.

Michael Leaf is the Director of the Centre for Southeast Asia Research (CSEAR) at the University of British Columbia (UBC), Associate Professor in the UBC School of Community and Regional Planning and Research Associate of the UBC Centre for Human Settlements. The focus of his research and teaching has been on urbanization and planning in cities of developing countries, with particular interest in Asian cities. Since his original doctoral research (PhD University of California, Berkeley, 1992) on land development in Jakarta, Leaf has been extensively involved in urbanization research and capacity building projects in Indonesia, Vietnam, Thailand, China and Sri Lanka. The courses he teaches cover the theory and practices of development planning and the social, institutional and environmental aspects of urbanization in developing countries.

Carolyn Loeb is Associate Professor of art and architectural history in the Residential College in the Arts and Humanities at Michigan State University. Her current research focuses on postwar and post-reunification redevelopment in Berlin. She has published on public sculpture (*Journal of Urban History*, 2009) and West Berlin murals (*Public Art Dialogue*, 2014) as these embodied and contributed to grassroots concerns about the history and future of the city. Other work has addressed planning history (*Planning Perspectives*, 2006) and, in edited volumes, cultural memory in the context of border redevelopment (2013, co-authored with Andreas Luescher) and housing in East and West Berlin (2013). She authored *Entrepreneurial Vernacular: Developers' Subdivisions in the 1920s* (The Johns Hopkins University Press, 2001).

Andreas Luescher is a Swiss architect, artist, and writer who is currently Professor and Graduate Director of Architecture and Environmental Design at Bowling Green State University, Ohio. He has written more than 60 papers for presentation at national and international conferences and for publication in leading international academic journals such as *The Journal of Architecture, Journal of Design Research, International Journal of Art and Design Education*, and *ArchNet-IJAR: International Journal of Architectural Research*. His first book, *The Architect's Portfolio* (2010), has been translated into Chinese. His second book, *The Architect's Guide to Effective Self-Presentation*, was published in 2014. He has authored two chapters published by Ashgate Publishing, and edited a special issue for *Urban Design International* entitled "Shrinking Cities and Towns: Challenges and Responses."

Conor McGrady is Dean of Academic Affairs at Burren College of Art, a school committed to an alternative model of art education. As an artist his work examines the relationship between ideology and the politics of spatial control. He received a BA Hons from the University of Northumbria and his MFA from The School of the Art Institute of Chicago (1998). He has exhibited internationally, with solo exhibitions in New York, Miami, Atlanta, Chicago and Zagreb. Group exhibitions include the 2002 Whitney Biennial in New York and the Biennale of Contemporary Art, D-0 Ark Underground, Sarajevo-Konjic, Bosnia and Herzegovina. Editor of "Curated Spaces" in *Radical History Review*, his writing has appeared in *The Brooklyn Rail, Ruminations on Violence* (Waveland Press, 2007), *State of Emergence* (Plottner Verlag, 2011), and *State in Time* (Drustvo NSK Informativni Center, Ljubljana, 2012). He lives and works in New York City and the Burren, Ireland.

David Monteyne is Associate Professor and Associate Dean of Architecture in the Faculty of Environmental Design at the University of Calgary. He teaches courses in the history and theory of architecture and urbanism. Monteyne has held fellowships at the Centre for Research in the Arts, Social Sciences and Humanities and Clare Hall, University of Cambridge, and at the Canadian Centre for Architecture, Montreal. He is the author of *Fallout Shelter: Designing for Civil Defense in the Cold War* (University of Minnesota Press, 2011).

Garth Myers is the Paul E. Raether Distinguished Professor of Urban International Studies and Director of the Urban Studies Program at Trinity College. Myers is the author of three books: *Verandahs of Power: Colonialism and Space in Urban Africa* (Syracuse, 2003), *Disposable Cities: Garbage, Governance and Sustainable Development in Urban Africa* (Ashgate Publishing, 2005), and *African Cities: Alternative Visions of Urban Theory and Practice* (Zed, 2011). He has co-edited *Cities in Contemporary Africa* (with Martin J. Murray; Palgrave, 2006, reissued in paperback, 2011) and *Environment at the Margins: Literary Studies and Environmental Studies in Africa* (with Byron Caminero-Santangelo; Ohio, 2011). Myers has also authored or co-authored more than 50 journal articles and book chapters.

Antonio Petrov is Assistant Professor at the University of Texas, San Antonio, and Caudill Visiting Critic at Rice University. He is program co-director of the *Expander* at Archeworks, Chicago, co-founder of the Harvard Graduate School of Design publication *New Geographies*, was a founder and is editor-in-chief of *DOMA*, and directs WAS, a Chicago-based think-tank. Major research foci include processes of urban and regional restructuring, questions of architectural articulation and how architecture as an expanded and geographically-inspired idea structures, shapes and produces complex territories. He recently edited volume 5 of *New Geographies*, "The Mediterranean: Worlds, Regions, Cities, and Architectures," and is currently working on a manuscript entitled "Between Autonomy and Total Immersion," which traces new forms of the secular in evangelical architecture in the United States. Petrov received his doctoral degree in the history and theory of architecture, urbanism, and cultural studies from Harvard University.

Kristin Poling is Assistant Professor of German history at University of Michigan Dearborn. She received her PhD in History at Harvard University (2011), and was Visiting Assistant Professor at the University of Rochester 2012–14. Her areas of teaching and research include comparative urban studies and the cultural and social history of the modern European city. Her current research focuses on how the idea of the frontier organized nineteenth-century understandings of space and progress in city development and the relationship between urbanism and nationalism. She recently published an article on Berlin's nineteenth-century shantytown frontier, and she is currently working on her first monograph on urban space and German national identity through the early twentieth century.

Tina Potočnik graduated in history of art from the Faculty of Arts, University of Ljubljana (2007) and received her master's degree in conservation and museology at the same faculty (2010). She concluded her studies with a doctoral dissertation at the Faculty of Architecture, University of Ljubljana (2013). Her research focuses on twentieth-century architecture, especially that of post-war Yugoslavia, with special emphasis on the work of architect Edvard Ravnikar, in the context of contemporary politics, ideology, identity, and the built heritage. She has been awarded several scholarships, including the Austrian government OEAD scholarship and the *Keepers Preservation Education Fund Fellowship*. She was assistant researcher at the France Stele Institute of Art History, Scientific Research Centre of the Slovenian Academy of Sciences and Arts, Ljubljana (2009–13). In 2013 Potočnik co-founded RI19+, Research Institute for Visual Culture from the Nineteenth Century to the Present Time in Ljubljana, where she works as an independent researcher.

Julia Walker is Assistant Professor of art history at Binghamton University, and has also taught at the Savannah College of Art and Design and at Yonsei University in Seoul. She received her MA and PhD in art history from the University of Pennsylvania. Her current book project, *Allegories of the Avant-Garde*, builds on the essay in this volume while broadening the theme to examine the ways in which Berlin's post-Wall building boom resurrects and reformulates architectural forms and theories from the early 1900s. Walker is co-editor (with Pepper Stetler) of an issue of *The Journal of Architecture* that investigates the use of the book to document, describe, promote, and critique modern architecture. She has published essays on Daniel Libeskind, David Chipperfield, and Zaha Hadid, among others. Her work has been supported by the Deutscher Akademischer Austauschdienst and the Society of Architectural Historians.

Editors' Acknowledgements

This book, a collaborative and interdisciplinary effort, would not have been possible without the wealth of knowledge and experiences generously shared with us by its contributors. Their commitment to this project inspired our own efforts.

We also would like to recognize those who went out of their way to help us during the process of research, in particular Petra Albrecht, Akademie der Künste, Berlin; Alexander Gall, Deutsches Museum, Munich; Alessandro Cattunar, University of Padua; Mari Fujita, University of British Columbia, Vancouver; Matthias Schirren, University Kaiserslautern, Germany; Wolfgang Voigt, Deutsches Architekturmuseum, Frankfurt; Wilko Hoffmann, J. MAYER H. und Partner, Architekten, Berlin; and Claire Dilworth, Dia Art Foundation, New York.

We extend our thanks, too, to the Society of Architectural Historians and to Abigail A. Van Slyck, General Chair of the 65th Annual Conference in Detroit in 2012, for providing the original venue at which some of the ideas presented in this book were first publicly explored.

We owe many thanks to the editorial expertise provided by Ashgate, particularly that of Valerie Rose, Publisher; Aimée Feenan, Senior Editor; Charlotte Edwards, Assistant Editor; and Matthew Carmona, Series Editor for *Design in the Built Environment*.

Finally, we wish to thank our families for their patience and solidarity.

Introduction: The Dialectics of Borders

Carolyn Loeb and Andreas Luescher

In a globalizing world, frontiers may be in flux but they remain as significant as ever. New borders are established—as they were recently to create the new state of South Sudan—even as old borders, such as that between East and West Germany, are erased. Frontiers emerge, shift, and disappear in response to varieties of historical, political, economic, social, and cultural exigencies.

Shifts like these represent the paradox of borders. While we may assume that some "natural" distinction between "here" and "there" creates a border, a line of division, it is often the creation of a boundary that charges such a fixed dualism with meaning. Once defined, a frontier can bring to the fore real or perceived differences in world view, products, ideas, and culture whose interchange is then facilitated or blocked. Furthermore, out of undifferentiated space, entities are produced that, once a border is defined, suggest other configurations and placements, perhaps stretching the imagination to change conceptions of what a border is. Although borders may persist for long periods, as long as they endure, their fixity remains in constant tension with alternative arrangements. Rather than settled constructions, borders often require adjustment, which suggests further ways in which they may be provisional, and this can lead to on-going interrogation of their form. Are the discriminating features completely defined? Have elements been left out? Are there further distinctions to draw? Are the burdens of separation foreclosing desired interactions? What is the impact of activities that are provoked by the stage that the frontier creates? Borders come into existence already carrying precedents and implications for their change or erasure.

This paradox at the heart of the idea of a frontier is itself a horizon that serves as a border for the chapters in this volume. The authors of the contributions collected here problematize the idea of borders by reflecting on how the tactics and techniques sensuously embodied in built form and environmental design intervene to shape meanings that are expressive of the relationships, practices, and tensions embedded in frontier sites. They examine spatial zones in which distinctive architectural, graphic, and other design elements are deployed to signal the nature of the space and to guide, if not actually control, experiences and behavior within it. The chapters in this volume unpack how manipulations of space and design in frontier zones, historically as well as today, set the stage for specific kinds of interactions and how they convey meanings about these sites, the experiences that the sites help to shape, and the nature of borders.

The artwork that Minimalist artist Walter De Maria created in California's Mojave Desert in 1968, *Mile Long Drawing*, can help us to see the power that visual media have to illuminate the paradoxical character of a border (Figure I.1). Against the uniformly unrolling scrub of the open desert landscape, the clean and visible marks he drew differentiated both themselves from their surroundings and the parcels of land they created from each other; plots of land were defined—the one on the right, the one on the left; here, rather than two yards further over there—and became potentially measurable.

At the same time, as a work of Land or Earth Art, a relatively temporary intervention in the on-going cycles of nature that construct this landscape, the marks or borders were fated to disappear. The decisive mark and its transitoriness exist in dialectical interplay and tension. The chapters in Part I of this volume, "The Border as a Line through Space," examine elaborations of this primal border mark and its implications, in the form of walls and fences.

The interdisciplinary chapters in this volume, then, bring a particular syntax to considerations of borders that enlarges the frame in which they are usually discussed. The authors draw on their expertise as urban historians, art and architectural historians, political scientists, urban geographers, literary scholars, and artists to focus, to varying degrees, on how spatial and architectural design decisions convey meaning sensuously, shape or abet specific practices, and stage memories of former frontier zones. In bringing such analyses together, this volume broadens and complements existing scholarship in the field of Border Studies and related work on themes such as divided cities (Wasti-Walter 2011; Ward 2011; Calame, et al. 2009; Verheyen 2008).

This project began as a co-chaired session at the 2012 annual conference of the Society of Architectural Historians. Held in Detroit, we thought that our topic, "Frontiers: Topographies of Surveillance and Flows," would elicit paper proposals on this city's status as a border site. None were, in fact, offered, but border city phenomena that resonate with the Detroit/Windsor frontier, such as how links and entrances—connectors and markers of separate identities—are defined, appeared as topics in relation to other locales. In addition, the range and quality of the proposals that we received, combined with the lively response to the session during the conference, suggested the idea of creating a volume of essays on this theme. Four of the contributions contained here were first presented at that conference session (Guidicini, Monteyne, Potocnik, and Walker), and two developed from abstracts that had been submitted to us. Other contributors were sought out on the basis of the work they were engaged in or, in some cases, were serendipitously encountered and discovered to share similar concerns. We are grateful to the SAH for the opportunity to explore this topic at the conference, to

I.1 Walter De Maria, *Mile Long Drawing* 1968. Mojave Desert, California. Courtesy of the Estate of Walter De Maria.

those who participated in and attended the session, and to all of the contributors to this volume for their commitment to this project.

The subjects, perspectives, and approaches that our authors address are varied, lending themselves to multiple forms of organization to highlight their numerous points of contrast and comparison. We have organized the volume into three parts that emphasize these scholars' architectural, spatial, and design concerns, which constitute the central focus of our inquiry. These parts—"The Border as a Line through Space," "Border Buildings," and "Spatial Ambiguity and (Dis)embodied Memory"—move us from the more concrete and defined to the more abstract and problematic, but they also introduce early on the kind of challenge to the concept of the frontier that, as we have seen, even as seemingly uncomplicated a statement as a line is capable of provoking. To enlarge the context for these chapters and the concerns they address, we include in our overview here examples of significant analyses drawn from scholars and design practitioners whose work is not included in this volume, as well as references to historical and contemporary situations on the ground that highlight the dialectical forces contending at borders.

THE BORDER AS A LINE THROUGH SPACE

The most basic way to define a border, as we have seen, is to draw a line through space; the chapters in Part I, "The Border as a Line through Space," address some of the many ways in which this can be construed in practice. In Chapter 1, Conor McGrady's, "Divison and Enclosure: Frankie Quinn's Peaceline Panorama Photographs" and its accompanying photos, begins our collection by underscoring the decisiveness with which walls and fences can divide space, in this case through the heart of the city of Belfast. These interface areas along the peacelines, as the shared boundaries along the walls are called, mark what have become unbridgeable distances and create "networks of enclaves, ghettos and deeply divided communities across the city," as McGrady tells us. Unyielding and inflexible as they are, as surfaces they nevertheless invite public if anonymous commentary, as seen in Quinn's photo of the graffiti, "Is there a hidden agenda." Nothing could be less hidden than the peacelines; the division is in "plain sight." But the walls and fences that define the boundary here in Belfast render even more obscure those on the other side, who become that much more easily always "other." And peace itself, if that is the agenda, remains hidden as well.

The unavoidable and severe presence of the walls and fences that divide Belfast can be contrasted with the private, symbolic enclosures, *eruvim*, used by Orthodox Jews to facilitate movement through and activities within public space on the Sabbath, which David Rotenstein has documented in published work not included here (Rotenstein 2011). Supported by natural and man-made topographical features, such as telephone poles and wires, these enclosures are defined by the additions of thin moldings or posts and string to represent the vertical and horizontal elements of a structure. These elements transform secular subdivisons, commercial streets, and even farms into a sacred space of which only members of the local congregation, all of whose homes are included within it, are aware. In addition to creating a boundary that is neither apparent nor operative for people in the wider community, the shape of an *eruv* can change, expanding to accommodate newcomers' houses or shrinking when members move away. Without interfering with secular functions, *eruvim* layer a charged, sacred space within the space of quotidian life by deploying a line as a border. Such a private world and its meanings, enacted in public space, contrasts with the brutal rupture in public space imposed by Belfast's stark walls.

Social geographer Alastair Bonnett recounts another relatively rare situation in which co-existence thrives amid a landscape fragmented by enclaves (Bonnett 2014, 215–22). Along the Belgian-Dutch border, the vestiges of medieval "micro-borders" survive in two villages; the Dutch town of Baarle-Nassau includes 22 Belgian enclaves, which comprise Baarle-Hertog, while eight bits of Baarle-Nassau are found within the Belgian enclaves. White crosses mark some pavements to indicate borders between the two nations, and house number signs sometimes include national flags, but there are so many borders within this patchwork community that marking them all has proven impossible. Nevertheless, the borders exist and everyone knows to which country they belong. At this point in history, and with both nations, as Bonnett points out, members of the European Union, there is little reason to attempt to refashion and rationalize this fragmentation. Rather, for Bonnett, this jumble of enclaves attests to the inspiring character of borders and to the paradox that while sometimes "they close down free movement yet [they also] suggest a world of choices and possibility" (Bonnett 2014, 216).

Security, surveillance, separation: these more frequent functions of boundaries that take the form of walls and fences can be identified at other sites, from the West Bank to the US–Mexican border. The triple layers of wire fences topped with razor wire at the Spanish enclave of Melilla within Morocco vividly express the sense of crisis felt by this outpost of the European Union in the face of the desperation of African migrants searching for relief from unemployment and conflict. The reinforced fence at the US–Mexico border is the subject of another noteworthy graphic consideration of the impact of these structures on communities and their development that is not part of this volume, "Radicalizing the Local: 60 linear miles of transborder urban conflict," a 2008 project by architect and social change activist Teddy Cruz (Estudio Teddy Cruz 2008). This border installation recalls the structure of the Berlin Wall memorial at Bernauer Strasse, an artful screen of vertical rods that alludes to the Wall's rebar but dematerializes its concrete mass. The more closely set and higher bars at the US–Mexico border between San Diego and Tijuana similarly toy with transparency but ultimately are effective barriers to movement across the frontier. And, as Cruz's study shows, over a 60-mile stretch the wall disrupts natural environments and habitats and distorts social relations as these are expressed through juxtapositions of gated communities and slums on either side of the barrier.

An earlier chapter of the Berlin Wall's history is the subject of Kristin Poling's essay in Chapter 2, "Occupying No Man's Land in the Lenné Triangle: Space, Spectacle and Politics in the Shadow of the Berlin Wall." The 1988 incident that Poling narrates exposed both the pragmatic arrangements for management of the Wall that the governments of East and West Berlin negotiated and the theatricality of the Wall as a public site, at least on the West Berlin side. Fifteen months before the Wall fell, environmental activists and those opposed to highway construction, with its associated demolition of the old urban fabric, took over a small piece of land that technically belonged in the East but was about to be officially transferred to Western control. This action signaled that however contested this space was, its location at the Wall no longer made it a flashpoint of the Cold War. Rather, by serving as a stage for the politics of local West Berlin urban development and by highlighting the marginal nature of neighborhoods located close to the Wall, the Lenné Triangle incident rehearsed some of the tensions that would surface in the eventual process of reunifying the city. At the time, however, the challenge represented by protesters' leap over the Wall to safety in the East upended the border's definitions of inside and outside and its regimes of security, surveillance, and separation.

The pragmatic considerations that rendered large and small areas of land exchangeable across the divide of the Berlin Wall—the Lenné Triangle is just one late and, in the event, rather spectacular example of this—belie the claim that borders make to mark necessarily inherent and decisive spatial difference. Dissolutions of borders, of course, challenge this claim as well. The creation of one large, borderless Europe through the Schengen arrangement in 1995 has raised questions about the nature of national, as well as European, identity that have only intensified as this zone has expanded over time. And practical issues have arisen as well that create murky situations along its edges. The Bosnian city of Neum, for example, provides Bosnia-Herzegovina's only access to the sea, interrupting the Croatian coastline for nine kilometers. Since Croatia joined the European Union in 2013, the national border here became a European Union border as well, creating additional concerns. To prevent goods from Bosnia-Herzegovinia that do not meet EU standards from entering the European Union through the Neum corridor and to avoid the need for additional Croatian border controls, an agreement was reached whereby goods passing through this border must be transported in closed vehicles identified with a seal and they must cross the corridor within a specified time period. This solution virtually decentralizes the border, distributing it among the closed vehicles that bear its seal; Max Hirsch recounts a similar resolution to the issue of transporting travelers from mainland China to Hong Kong International Airport that is described in the next section.

The mutual reinforcement, dependency, and precariousness that borders in the form of gated communities cultivate between elites and the poor, identified in Teddy Cruz's project mentioned above, are also the focus of Garth Myers' contribution in Chapter 3, "Remaking the Edges: Surveillance and Flows in Sub-Saharan Africa's New Suburbs." By pointing to paradoxical affinities that belie the divisions borders are intended to mark, this chapter foreshadows, too, the attention to the instability and arbitrariness of the line drawn through space that the authors in the third part of this volume address. The surveilled, bounded enclaves of corporate business parks and housing developments in Lusaka, Zambia (formerly Northern Rhodesia) that Myers analyzes are simultaneously partner to and antithesis of the unconstrained movement of money, materials, and personnel required by the international investors who build them. Furthermore, these enclaves share the carceral character of the informal, marginal settlements that make up the expanding urban periphery. The intentional, designed disconnection and invisibility of gated communities, produced through the erection of walls as well as screens of exotic plantings, are uncannily echoed by means of the arguably intentional withdrawal of design from marginal settlements, which are characterized by a barren lack of trees and other plants, isolation from urban transport and pedestrian routes, and absence of street names or signage that render marginal settlements equally invisible.

The point along a border at which the distinction between inside and outside is negotiated and at which the possibility for transformation occurs is at its opening; entranceways affirm the need for formal admittance into a bordered space and thereby also deny its impermeability. In Chapter 4, "Imagining and Staging an Urban Border: The Role of the Netherbow Gate in Early Modern Edinburgh,"

Giovanna Guidicini examines the variable meanings of the frontier between the city of Edinburgh, with its commercial, defensive, and administrative privileges and obligations, and the religious and royal service community of Canongate. Netherbow Gate exemplifies the function of the border to demarcate the realm of order and law from a diffuse space of lawlessness and potential chaos. As space becomes redefined over time, borders shift, leading even such solid structures as gates to lose their purpose; Edinburgh's annexation of Canongate in 1636 foreshadowed Netherbow Gate's destruction. And yet the liminal power of border gateways persists in urban space, memorialized at the least in place names if not in built form.

It is not uncommon for traces of borders to remain to memorialize earlier frontiers. The Great Wall of China is perhaps the most massive and widely known example of this. Less monumental markers such as street patterns can hold memories of previous boundaries, too, such as Vienna's Ringstrasse, which encircles the city as had the earlier *glacis*, the open area that permitted surveillance in front of the city wall. A more contemporary example of spatial memorialization can be found in the European Greenbelt Initiative, which aims to preserve the zone that marked the divide between the European Eastern and Western blocs during the Cold War as a natural area devoted to the promotion of ecologically sustainable development.

BORDER BUILDINGS

In Part II, "Border Buildings," examinations of more complex structures and sequences of passage that mark frontier experiences elaborate on the role of the gate to which Guidicini introduces us. Perhaps the clearest, if most extreme, relationship between a building and a border is that represented by an embassy. Here, the physical structure is politically and symbolically identified as territory that is otherwise linked only notionally to a geographical location that may in fact be half a world away. Routine public services that an embassy provides, such as visa processing and distributing information about the given country, can communicate features of its culture; but the distinctive nature of the embassy as extra-territorial is highlighted especially in political contexts when, for example, dissidents or others seek sanctuary within it, beyond local borders and within the frontiers of a safe haven. Such a conceptual superimposition of the embassy's space on that of the host country exposes the malleability of defining spatial entities that otherwise may be unsuspected or denied.

The Haskell Free Library and Opera House represents a challenge of a different sort to the definition of a frontier space. This building, completed in 1904, was deliberately situated directly on the border between Stanstead, Quebec, and Derby Line, Vermont, to provide cultural benefits to citizens of both Canada and the US within the same structure. The building is not extra-territorial; access is dependent on good relations between the two governments and perhaps, too, on the institution's out-of-the-way site. Nevertheless, one entrance lies in the US, another in Canada, and the two citizenries mingle as they unconsciously move back and forth across the border to check out a book or attend a performance. The building refuses the separation that the border defines (Haskell 2014).

In Chapter 5, Richard Kurdiovsky examines a more complicated relationship between a structure and bordered space. In "House Number 1: The Vienna Hofburg's Multiple Borders," Kurdiovsky analyzes how the palace of the Hapsburg king and Holy Roman Emperor, a concatenation of building wings and courtyard spaces that evolved and shifted over time, created border situations within the context of everyday Viennese urban life. The Hofburg channeled a flow of daily movement into and from the heart of the city, much as Netherbow Gate did, while it also enforced the boundaries of monarchical space. Physical integration of these royal buildings and spaces within the city posed challenges to the necessity for separation between social classes, on the one hand and, on the other hand, offered the monarch an opportunity to create a unique rapport with his subjects based on the quotidian, *burgerlich* proximity that the numbering of the palace as House Number 1 quaintly suggests. As a set of structures, the Hofburg articulates interpenetrating spaces and distinctions between impermeable and crossable boundaries that accommodate its urban situation and that coincidentally provide a handbook of architectural solutions for building multi-functionally and multi-dimensionally in an urban setting.

The succeeding two chapters address twentieth- and twenty-first-century national border installations. Not only do national borders themselves shift, but the experience of border-crossing can be shaped in response to security concerns, national self-definition and how this is projected, expected types and situations of border-crossers and how they are viewed, among other issues. Pier 21 in Halifax, Nova Scotia, the subject of David Monteyne's chapter, "Pier 21 and the Production of Canadian Immigration," was operational as an entry-point from 1928 to 1971. Monteyne penetrates its banal, warehouse-like exterior, with its suggestion that the immigration process inside would proceed with assembly-line precision, to recover what he describes as the "layered and interwoven, even chaotic" character of the experiences within. Using oral histories, Monteyne reanimates the "performed space" in which immigrants speaking many different languages and from many different walks of life, energetic and curious children, social workers, nurses, volunteer Canadian citizens, customs officials, and others mingled. Against the backdrop of the legal and architectural grid of order—procedures, benches, closed doors, baggage cages—flowed the unpredictable and uncontainable activities of the people who passed through it. Offering none of the symbolic portent of the entry to the United States at Ellis Island, Monteyne argues that Pier 21 presented a more focused experience; it indicated nothing about the new lives awaiting them, but from the train spur alongside it, Pier 21 sped immigrants toward it.

Eric Aronoff and Yael Aronoff present another close reading of border installations in Chapter 7, "Bordering on Peace: Spatial Narratives of Border Crossings between Israel, Jordan and Egypt." Through analyses of the spatial organization, signage, and visual representations found in installations at four crossings, they locate the programmatic, differential meanings that each projects. Designed and administered by the Israel Airports Authority, it seems to be neither accidental nor especially intentional that the Israeli installations suggest airport terminals. This speaks to a general feature of contemporary borders, which lack a fixed vocabulary to embody the variable meanings of frontiers; as a result, they take such forms as walls, highways, green spaces, or warehouses, as previous chapters indicate, as well as airports, as the Aronoffs recount, or even highway toll plazas, as at crossings between Canada and the United States. Even sealed vehicles can serve this function, as in the Neum corridor discussed above or as described by Max Hirsh in his study of the infrastructure created to allow undocumented travelers or those with modest means to cross from mainland China to Hong Kong International Airport without going through customs and immigration controls (Hirsh 2014). The liminal border space of international airports is also affirmed by urban legends about people forced to live there for years and the actual cases of undocumented travelers' detention in airports on which they are based; the enforcement of frontiers entails definitions of exclusion that in some cases, such as those of stateless individuals, can lock people at the very border that seeks to exclude them.

A rare and dramatic contrast to the anonymity and bureaucratic rationalization that mark many installations is found in the 2011 checkpoint at the Sarpi, Georgia, border with Turkey by the German architectural firm J. MAYER H. (Figure I.2). Taking its dynamic silhouette from the coastline it punctuates, architect Mayer H. observed of it that: "If a country says hello and goodbye with a building like this it's a gesture demonstrating that country's openness to the world" (Dvir 2014, 70). With conference rooms and terraces located in its tower overlooking the sea, the border installation here is interpreted as a site of connection rather than separation.

The differences among the crossings that the Aronoffs register in the "narratives of relation" they tell through imagery and signage as much as through architecture reflect the complex histories and nuanced border arrangements between these countries. Israel's borders with Jordan and Egypt are relatively recent achievements that contribute to Israel's not uncontested status as a legal participant in the community of nations. Against a backdrop of vulnerability, the narratives of the peace process and of the Hashemite monarchy's line of succession that visitors encounter at the Itzhak Rabin and Wadi Araba crossings between Israel and Jordan, respectively, nevertheless make claims to stability and continuity. At the Allenby Bridge/King Hussein crossing, both address a primarily Palestinian population making the crossing, Israel by focusing on the Islamic sites of pilgrimage in Jerusalem to which many of these visitors are bound, and Jordan by highlighting the Hashemite royal family and especially Queen Raina, who is herself Palestinian. These relationships are made meaningful for specific sites and the populations using them through the organization of spatial experience at the borders.

Julia Walker's analysis of the design of the new German federal government center in Chapter 8, "The View from Above: Reading Reunified Berlin," brings forward to the contemporary period, but in a very different context, the ability of buildings to bridge and separate that Kurdiovsky discusses in relation to Vienna's Hofburg. Axel Schultes and Charlotte Franks' *Band des Bundes* (ribbon of federal buildings) at the bend in the River Spree (*Spreebogen*) between the inner city and

I.2 J. MAYER H. Architects, Border Checkpoint, Sarpi, Georgia. Photograph by Jesko M. Johnsson-Zahn.

its northern districts embodies the project of national unification functionally, but also through its deployment of differentiated levels of spatial connectors: underground passages, walkways on the ground plane, and both public and limited-access bridges between buildings and parks on both sides of the river. These articulate and, as Walker frames her analysis, allegorize the site on the border of East and West Berlin that the complex occupies. By allegory, a term used by the architects as well, Walker refers to the fragmentary, contingent, unresolved character of these structures, which allude to the ruptures and initiatives of the past even as they evoke desires for unity, transparency, and democracy that may not be fully achieved.

SPATIAL AMBIGUITY AND (DIS)EMBODIED MEMORY

Walker's focus on the project of dissolving a border to create a new unity whose permanence can be no more assured than to pose it as a desire that is always deferred also introduces themes addressed in Part III, "Spatial Ambiguity and (Dis)embodied Memory." Tina Potočnik's essay in Chapter 9, "Gorizia and Nova Gorica: One Town in Two European Countries," examines the sequence of historical shifts in the border between Italy and Slovenia. Whereas Gorizia was once a regional center, the city of Nova Gorica traces its origins to the Cold War division between the two countries that cut off Slovenians from the old city. Today, reconfigured once more in the wake of Slovenia's 2004 membership in the European Union and inclusion in the Schengen area in 2007, the two cities form a "transnational conurbation" that is a "space of open perception." Urban structure preserves the memory of the former frontier, however, as the margins of the cities that were shaped by the border now lie at the center of the conurbation; the site of unity memorializes the fissure of the past, creating a center-less center.

Efforts to acknowledge complexities of the past in design solutions that allow for both the preservation of memory and a way to build upon it that Walker's and Potočnik's chapters address can be contrasted with less successful transformations of former border sites. At the iconic Berlin Wall crossing-point of Checkpoint Charlie, for example, a pastiche of touristic elements provides no insight into the complex urban fabric and webs of human interactions that were associated with the site before and during the Wall's existence (Loeb and Luescher 2014). Original border installations have been removed, replaced with a small mock-up of a guard station that provides only a distorted sense of the spatial extent of the crossing. This is accompanied by stand-ins for border guards, with whom tourists can pose for photos. The dispersal of the site's history among an open-air exhibit chronicling Cold War history, erected by the municipal government, a private museum of memorabilia that focuses on attempts to escape through the Wall, and a panorama of the divided city by Berlin artist Yadegar Asisi, bombards visitors with imagery in a setting that has been emptied of all genuine historical references.

Ambiguity lies at the heart of the "desakota" zones that Michael Leaf analyzes in Chapter 10, "New Urban Frontiers: Periurbanization and (Re)territorialization in Southeast Asia." These borders between rural and urban areas highlight frontiers as a discourse and as zones of encounter, contestation, change, and uncertainty. They are tropes for development itself as a frontier of globalization, where flows of foreign investment, land speculation, and media and marketing campaigns meet "state intentionality" exercised, or not, through property and regulatory controls. Leaf's case in point, real estate development in Ho Chi Minh City, juxtaposes globalization of the luxury residential market with widespread local involvement in housing provision for migrant workers. Bringing small landowners into the arena of real estate development, even on such a modest scale, helps to build decentralized governmental systems within a complex and flexible fabric of local, national, and international relationships that are themselves emblematic of the fluidity of borders.

The Mediterranean basin's historically shifting frontiers are metaphors for boundaries' ambiguities and for their dissolution and preservation through memory. Such "liquid geographies" are the subject of Antonio Petrov's essay in Chapter 11, "Mediterranean Frontiers: Ontology of a Bounded Space in Crisis." Petrov considers two utopian—or possibly dystopian—proposals for reconfiguring borders that destabilize notions of bounded space, Herman Sörgel's *Atlantropa* from the 1930s and Rem Koolhaas' more recent plan for *Eneropa*. Both take technology, and especially the need to develop viable energy sources, as their inspiration for "overwriting" existing borders. Related examples of instances in which regions transcend borders exist on the ground today, as in "Cascadia," the "bioregion" in which the western states of Canada and the US coordinate environmental reforms, or in "Regio TriRhena," the collaboration among France, Switzerland, and Germany that created and administers the EuroAirport Basel-Mulhouse-Freiburg. A similar proposal has been made to allow travelers access to Tijuana's International Airport through a passageway through the border between the US and Mexico. As scholar Fernando Romero notes, such cross-border collaborations often address needs more successfully than individual national governments do (Romero 2008, 33). An older but more quixotic instance is the condominium, or joint administration, of Pheasant Island, located on the French-Spanish border in the middle of the Bidasoa River. Since the 1659 Treaty of the Pyrenees, sovereignty over this site has alternated every six months between France and Spain (Jacobs 2012). Broadening our range of references, we may consider not only the European Union—itself once considered utopian—as a version of the proposals Petrov presents, but also the spatial implications of the United Nations, its Universal Declaration of Human Rights, and UNESCO's World Heritage Conventions. All of these examples and the projects discussed by Petrov open the door to yet-unimagined ways to reconceptualize global space and the status of borders.

Indeed, through the force of experience and the proliferation of new needs, or through prescient design proposals, pressure can be placed on existing frontier arrangements, which eventually give way to solutions that are more responsive to contemporary attitudes and ways of living. Many current presuppositions about and infrastructures for the construction of borders today—highway barriers, mixed-use enclaves, and gated communities of the sort discussed by Myers and Leaf, for example—rely on insights first realized on the imaginative plane by Norman Bel Geddes in his Futurama installation for the 1939 World's Fair in New York City (Albrecht 2012). Sponsored by General Motors and seen by almost five million people, Futurama highlighted the transformative role that automobiles would come to play in reshaping cities as infrastructure-dependent engines of mass production and consumption. As this vision continues to play out in the twenty-first century, the landscapes of isolated, bordered commercial or residential neighborhoods held together by feeder roads that it underwrites increasingly exist in tension with shrinking energy resources, needs for community, and desires for a sense of place.

The open-endedness of the proposals that Petrov presents and their encouragement to rethink the nature of frontiers nevertheless tend to run up against continuing legacies of border formation, especially those based on definitions of the nation-state. The virtual *State in Time*, created by the contemporary Slovenian conceptual artists Neue Slowenische Kunst (NSK), confronts this tension, as Conor McGrady makes clear in the final chapter, "The NSK State and the Collective Imaginary," that accompanies images of their work. A project that embodies the artists' "virtual secession" from Slovenia in 1992, when the region was wracked by nationalist conflicts, it represents a de-territorialized "utopian social sculpture" beyond "nations, boundaries and territories." Yet the NSK

State's passports, embassies, and military are haunted by and engage the dystopian aspects of nation-states, especially their authoritarianism. The artists use visual media and performance to present a context in which to "test new ways of living"; this aim links their work to the Occupy movement, with which it resonates. The artists offer us the opportunity to examine our conventional imaginaries, which have been formed within the confines of bounded political and social structures, and to experiment with alternatives in the freedom of virtual space.

All of the visual representations of bordered space discussed and collected here, whether built or speculative, point to the inherent dialectic between fixity and fluidity that characterizes frontiers. To assess the challenges of globalization as well as its opportunities in relation to humanity's increasing precariousness, on the one hand, and its expanded realms of freedom, on the other, it is useful to consider the implications of the many struggles underway today for and against boundedness. The question of when or whether new frontiers provide the security and ordered environment within which people can flourish, or whether new arrangements that emphasize flexibility, multiplicity, and movement provide alternative safeguards that promote full human existences, must be considered against the backdrop of past and existing frontier dynamics, including the meanings revealed by their designed forms.

REFERENCES

Albrecht, D. (2012). *Norman Bel Geddes Designs America*. New York: Abrams.

Bonnett, A. (2014). *Off the Map: Lost Spaces, Invisible Cities, Forgotten Islands, Feral Places, and What They Tell Us About the World*. London: Aurum Press.

Calame, J., Charlesworth E., and Woods, L. (2009). *Divided Cities: Belfast, Beirut, Jerusalem, Mostar, and Nicosia*. Philadelphia: University of Pennsylvania Press.

Dvir, N. (2014). "The Georgian Miracle," *Art in America*, March: 67–70.

Estudio, T.C. (2008). "Radicalizing the Local: 60 linear miles of transborder urban conflict," http://www.moma.org/collection/object.php?object_id=126222 (video available at: http://estudioteddycruz.com/index1.html#!prettyPhoto/4/).

Haskell Free Library and Opera House (2014). http://haskellopera.com/history/.

Hirsh, M. (2014). "Design Aesthetics of Transborder Infrastructure in the Pearl River Delta," *Journal of the Society of Architectural Historians*, 73:1, March.

Jacobs, F. (2012). "The World's Most Exclusive Condominium," *New York Times*, 23 January.

Loeb, C. and Luescher, A. (2013). "Cultural Memory after the fall of the Wall: The Case of Checkpoint Charlie," in Stephenson, M. and Zanotti, L. (eds) *Building Walls and Dissolving Borders*. Farnham: Ashgate Publishing, pp. 177–91.

Romero, F. (2008). *Hyper-Border: The Contemporary United States-Mexico Border and Its Future*. New York: Princeton Architectural Press.

Rotenstein, D. (2011). "Courtyards of Convenience: Montgomery County's *Eruvs*," *Vernacular Architecture Newsletter*, Summer.

Verheyen, D. (2008). *United City, Divided Memories?: Cold War Legacies in Contemporary Berlin*. Lanham, MD: Lexington Books.

Ward, J. (2011). *Post-Wall Berlin: Borders, Space and Identity*. London: Palgrave Macmillan.

Wasti-Walter, D. (2011). *The Ashgate Research Companion to Border Studies*. Farnham: Ashgate Publishing.

PART I
The Border as a Line through Space

1

Division and Enclosure: Frankie Quinn's Peaceline Panorama Photographs

Conor McGrady

In the wake of the Irish peace process, Belfast has faded from international news headlines in recent years, barring occasional interventions into the media spotlight upon sporadic, yet ongoing, eruptions of violence. Despite the perception that the conflict known euphemistically as "The Troubles" is over, deep divisions and simmering tensions continue to characterize cities like Belfast. Nowhere is this sense of ideological and political polarization more acutely felt than in the working class districts in the north, west and east of the city that are crisscrossed and interrupted by a series of monolithic walls and security barriers known, not without a sense of irony, as "Peacelines."

Twenty-plus years after the fall of the Berlin Wall, territorial division and geographical enclosure continue to proliferate. The fortification of the US-Mexican border parallels that of India and Pakistan, while Israel's architectural separation and division of the Palestinian West Bank highlights a politics of national and geopolitical entrenchment that shows no immediate sign of abatement. In the context of Northern Ireland, the visible manifestations of territorial control and separation are nothing new. Recent figures (Belfast Interface Project, 2011) show that Belfast's urban geography is carved up by no less than 99 peacelines, some of which have been in existence since the eruption of the recent conflict in the late 1960s. As the city fragmented along political and ideological fault lines, initially sparked by loyalist attacks on the nationalist lower Falls district, barricades were thrown up as defensive bulwarks. With the deployment of the British army, who sought to gain control of a city spiraling into the chaos of urban guerrilla warfare, the geography of separation became a means to control mobility within the insurgent nationalist community. Far from being a temporary measure, the barricades became walls that increased in number and in height over the years, forming a network of enclaves, ghettos and deeply divided communities across the city. In the Short Strand, a small nationalist area surrounded by the predominantly loyalist East Belfast, new walls continued to be erected and older ones extended up to heights of 40 feet in the first decade after the signing of the

Good-Friday Agreement. With tensions remaining high and division endemic, these walls and security barriers provide a prescient architectural testimony to a city in which approximately 94 percent of public housing is segregated along political and religious lines (Shuttleworth and Lloyd, 2007). In recent years the term "peaceline" has become interchangeable with "interface," denoting the common boundary between divided zones referred to as interface areas. In the wake of the peace-process groups such as the Belfast Interface Project monitor the impact of the social divisions in these areas and work to ameliorate tensions and support community regeneration. In his photos of the peacelines and interface areas, Belfast photographer Frankie Quinn not only captures the architecture of a divided city, but the impact of ideological polarization and topographical fragmentation on the lives of those who live alongside these structures. As Gabbi Murphy points out in her essay on this series of photos:

> *For all that the barriers may provide psychological reassurance for inhabitants on both sides, their security value is not absolute. A CAJ (Committee on the Administration of Justice) report of 2001 found that, while the erection of new barriers and closing of peaceline gateways in north Belfast reduced instances of criminal damage, they did not deflect rioting in those areas. Interface violence is still a fact of life, as is the anxiety of defensive living. Displays of culture and tradition, including parades and associated protests, still prompt escalations in tension and hostility. Perhaps even more lethal is the apathy that regards segregation as indelible. (Murphy, 2010, p. 178)*

Latent tension permeates many of these images. They embody the banality of disruption, where walls slice through housing developments and wire mesh imprisons as much as protects. In one image graffiti poses the question, "Is there a hidden agenda?" While the military apparatus has largely been dismantled in Belfast, security, policing and the politics of control continue to define the city. While its reference point evokes the specter of political control, the irony of the question addresses the divisions that exist in plain sight. Those divided remain hidden from each other. Marked as the perpetual "other," the sense of fear, suspicion and antagonism that characterizes such separation is ultimately tied to the overall fate of post conflict Northern Ireland.

Other images show the innocuous, almost banal, quality of daily life in the interface areas. Children's swings sit at the base of one wall, on which repeating decorative motifs attempt to ameliorate the reality of its core function. More often than not, an iconic, brutalist utilitarianism characterizes the walls, particularly those constructed at the height of the conflict. It is this language of the military fortification rather than the quasi-decorative attempts at creating structures that are "integrated" with the architecture of a particular area, that attract the tourists depicted in another image. Murphy (2010, p. 180) acknowledges the incongruous appeal of the more forceful signifiers of a divided city, which are now largely absent from the city center. She points out that "Quinn's pictures of camera-laden sightseers dipping their toes in the aftermath of conflict highlight the division between the buzzing hub of commercial regeneration and those who have been left behind."

While absurdity is a ubiquitous presence in these works, it remains dominated by the harsh geometry of urban control. In these images the language of force exerts constant pressure. The walls leave an indelible stamp on the lives of those living within their proximity, serving as a barometer of societal progress in the wake of the Peace Process. Their continued existence, which is regarded as necessary, if unfortunate, by many of those living in interface areas, highlights the disjuncture between political progress and the distances yet to be traveled in post-conflict Northern Ireland.

Frankie Quinn is a Belfast-based photographer. His interest in documentary photography developed as a result of his involvement with the MacAirt Camera Club in East Belfast. Since 1983 his work has been exhibited extensively both at home and abroad. His work has also appeared in numerous local publications including "Falls in Focus" published by the Falls Community Center (1987) and "Shoot Belfast" (1986), a guide for amateur photographers which was funded by the Northern Ireland Arts Council. His work has also appeared in the book "Garvaghy Road: A Community Under Siege" (1999). He was a founding member of the Belfast Exposed Community Photography Resource Center.

REFERENCES

Interfaces Map and Database (2011). *Belfast Interface Project* [online]. Available at: http://www.belfastinterfaceproject.org/interfaces-map-and-database-overview (accessed 4 October 2013).

Murphy, G. (2010). "Streets Apart: Photographs of the Belfast Peacelines," *Radical History Review*, 108, Fall.

Shuttleworth, I.G. and Lloyd, C.D. (2007). *Mapping Segregation on Belfast NIHE Housing Estates*. Northern Ireland Housing Executive [online]. Available at: http://www.nihe.gov.uk/mapping_segregation_final_report.pdf (accessed 4 October 2013).

1.1 Bombay Street, West Belfast.
Photograph © Frankie Quinn, Belfast, 2006.

1.2 Glenbryn/Ardoyne, North Belfast.
Photograph © Frankie Quinn, Belfast, 2006.

1.3 Glenbyrn/Alliance Avenue, North Belfast.
Photograph © Frankie Quinn, Belfast, 2006.

1.4 Mayo Park, West Belfast.
Photograph © Frankie Quinn, Belfast, 2006.

1.5 Lanark Way/Springfield Road, West Belfast.
Photograph © Frankie Quinn, Belfast, 2006.

1.6 Northumberland Street. Between Falls and Shankill Roads, West Belfast. Photograph © Frankie Quinn, Belfast, 2006.

1.7 Springmartin Road/Ballygomartin Road, West Belfast.
Photograph © Frankie Quinn, Belfast, 2006.

1.8 Torrens/Wyndham Street, North Belfast.
Photograph © Frankie Quinn, Belfast, 2006.

1.9 Springfield Road, West Belfast.
Photograph © Frankie Quinn, Belfast, 2006.

1.10 Springfield Road, West Belfast.
Photograph © Frankie Quinn, Belfast, 2006.

2

Occupying No Man's Land in the Lenné Triangle: Space, Spectacle, and Politics in the Shadow of the Berlin Wall

Kristin Poling

July 1, 1988 was the only date on which there was ever a mass flight from West Berlin into East Germany. On this day, a group of nearly 200 Kreuzberg protesters escaped West Berlin by leaping the Wall into the custody of East Berlin border guards. Their flight was the culmination of a six-week long standoff between West Berlin police and a small group of environmentalists and squatters who wished to protect a plot of land called the "Lenné Triangle," situated between the Wall and the western half-city's largest park, from the possible construction of a highway. That early July morning, the German Democratic Republic (GDR) welcomed the refugees with all the courtesy due to a public relations coup that had fallen into its lap, gave them breakfast, and sent them back home: for all the event's exceptionality, a rather unremarkable conclusion.

As a highway revolt turned Cold War incident, this singular event is often cited in histories of divided Berlin for its curiousness. In its inversion of the border's usual dynamic, in which refugees typically fled from East to West, it serves as an illustration of how the Berlin Wall and the border it marked rendered daily life and ordinary urban politics bizarre. Brian Ladd introduces it as "one of the oddest incidents in the Wall's history," using it to conclude a discussion of the "peculiar urban backwaters" and forbidden crossings in the Wall's shadow (1997, pp. 16, 13). Janet Ward refers to the same incident as a moment of "Schneideresque absurdity" (2011, p. 85). The reference is to Peter Schneider's classic treatment of divided Berlin's irrational urban geography in the novel *The Wall Jumpers* (*Der Mauerspringer*, 1982), which explores the ways in which the psychological effects of the city's division, though perverse, had by the early 1980s become utterly pervasive and virtually normalized. For Ward, the unexpected story reinforces a general lesson drawn from Schneider's book about human nature: "the firmer the barrier, the stronger becomes the human urge to conquer its law" (2011, p. 84).

In these accounts, the Lenné Triangle incident was simultaneously an exceptional event and just another entry in a long catalogue of events demonstrating that in divided Berlin the extraordinary was part of the everyday. It was what David Clay

Large calls "another object lesson in the bizarre theatricality" of protest politics in West Berlin (2000, p. 494). Cited as evidence of absurdity, the events of July 1, 1988 might even appear more as a game, or a farce, than as a serious event. In these narratives, the odd and the farcical serve symbolically to undermine the Wall itself and, implicitly, to anticipate its imminent collapse.

This chapter will examine this particular incident of unrest—so often mentioned in passing, but rarely closely considered—to examine how the Wall operated on a micro-level as a contested environment, arguing that at the time it did not seem to undermine the Wall and the international border it represented at all. Instead, it seemed to re-enact and reinforce West Berlin's peculiar status as both a frontier city and a landscape of multiplying borders and fortifications, reproducing the city's exclave spatiality and cultural politics in microcosm. The Berlin Wall provided a globally visible stage of spectacle and surveillance, investing local conflicts over the urban environment with a broader geopolitical significance (Pugh, 2010, p. 156). In the Lenné incident, both protesters and West Berlin city officials were aware of watching eyes, from East Germany, from West Germany, and from all over the world. At first, West Berlin's landscape of surveillance empowered marginalized Berliners by providing them with a widely visible stage for protest, but ultimately this same context defused the particular import of their local complaints by absorbing them into all-encompassing stories of Cold War conflict and Berlin's peculiar history. The way in which this occurred provides particular insight into how the Berlin Wall operated as a frontier space both locally and globally.

NESTED TRIANGLES: WEST BERLIN, KREUZBERG, LENNÉ

Whether described as "the ultimate postmodern space," a "total work of art" putting the free market on display, or a "surreal cage" (Borneman, 1992, p. 1; Steinfeld, 1990, p. 256; Taylor, 2006, p. 355), Berlin's exceptionality has been among the most persistent themes in treatments of its history and culture. In divided Berlin, the source of Berlin's exceptionality is evident, though some scholarship seeks to integrate this brief phase into a long history of the city's persistent marginality relative to Germany as a frontier capital, often with reference to Karl Scheffler's classic account from the beginning of the century (Scheffler, 1910; Webber, 2009, p. 2; Haxthausen and Suhr, 1990, p. xi). Between 1961 and 1989 West Berlin functioned both as a borderland and as a cultural and political island. As an exclave, it became alienated from the "mainland" of West Germany, as well as being isolated from its East German surroundings (Richie, 1998, p. 775). Residents enjoyed extensive economic subsidies and were exempt from military service, further distinguishing them from the rest of the population, and there was talk of an alternative "second culture" in opposition to mainstream West Germany. By the 1970s, "the very act of residing in West Berlin was considered a political statement" (Mayer, 1996, p. 217).

Being the furthest outpost of capitalist culture made West Berlin a natural showcase for western free market society. Much in the way East Germany pumped resources into the eastern half of the city because it was their most internationally visible urban area, the West German government heavily subsidized cultural activities in their outpost city. By the end of the 1980s, West Berlin supported an annual budget for cultural activities more than half that of the entire United States, but over half of the city's budget was supported by federal subsidies (Colomb, 2011, p. 52). West Berlin's famous scene "was a cultural program that served political ends" (Haxthausen and Suhr, 1990, p. xiv). In the most obvious and extreme cases, western culture was literally projected over the Berlin Wall in an attempt to gain converts in the East (Flemming and Koch, 2005, pp. 62–3). Even when less obviously aimed at conversion, a free and radical culture in West Berlin was transformed by its position on the front lines from suspect to politically important (Steinfeld, 1990, p. 254).

In a divided city defined by its proximity to the East, the neighborhood of Kreuzberg was the West Berlin city district the geography and environment of which were the most drastically shaped by the construction of the Wall. This working class district, located at the meeting point between the Soviet, British, and American sectors was surrounded by East Berlin on three sides after the construction of the Wall, making it a virtual island of Western culture jutting into "actually existing socialism." Although it had formerly been located in the heart of the city, just south of Berlin Mitte, most roads linking Kreuzberg to nearby cultural and commercial destinations were cut off by the Wall, or left in its traffic shadow (Lang, 1998, p. 111). In its relationship to West Berlin, Kreuzberg shared many of the traits that West Berlin had in its relationship to West Germany: a proud radicalism combined with cultural marginality that was created by physical inaccessibility and proximity to the border with East Germany.

Explanations of Kreuzberg's social and economic marginality drew on mutually reinforcing short- and long-term causes. Its position in the divided city together with its longer prewar past as a poor, working class district both contributed to Kreuzberg's economic struggles and social instability. The relationship between Kreuzberg and West Berlin had been antagonistic ever since the 1960s, when residents of the neighborhood had first protested massive renovation projects that would have obliterated much of the district's urban fabric (Lang, 1998; Karapin, 2007). In addition to poor housing stock and its isolated location, Kreuzberg was also cut off from most major parks (Lang, 1998, p. 37). Although it lacked officially designated green spaces, one of the most distinctive environments of Kreuzberg was the weedy and overgrown plot, particularly in the many forgotten corners where the Wall's construction had preserved empty spaces left behind by war damage.

Sociologist Barbara Lang suggests that both those who identified with the Kreuzberg environment and those who saw it as a threat depicted it as an environment that was cut off from and alien to the rest of West Berlin. A vocabulary of "outside" and "inside" created a rift between the border district and the rest of the border city. She writes that both sides acted as if "in Kreuzberg one lives behind an invisible line of demarcation" (Lang, 1998, pp. 134–6).[1] Lang suggests a link between the isolated and overgrown physical environments of Kreuzberg and the neighborhood's culture of radical protest. One of Lang's interviewees describes her fondness for the plot of land behind the Görlitzer Bahnhof: "*Wild* and overgrown was how it was somehow … I loved it so much" (Lang, 1998, pp. 195–6). The description echoes her account of her own life in a Kreuzberg squat where she was able to live as she would, without externally imposed controls or limitations. Whether with positive or negative connotation, Kreuzberg's location on the Wall, which ostensibly divided it only from the East, ended up separating it from West Berlin as well.

The sense of Kreuzberg's isolation from the rest of the city was worsened in the 1980s when a series of demonstrations and violent clashes between protesters and the police cemented the district's radical reputation. Opposition to the now right-center Christian Democratic Union (CDU) led city administration again peaked with renewed threats to return to the old policies of demolition and slum clearance insensitive to the neighborhood's own residents (Karapin, 2007, p. 94). Kreuzberg's reputation for violence became so extreme that in anticipation of agitation during President Ronald Reagan's visit to West Berlin in June 1987, shortly after the particularly violent unrest of May 1, 1987, Kreuzberg was simply closed off from the rest of the city. The subway lines into the neighborhood were closed and police were posted on streets around its edge. Describing the controversial tactics taken to protect the city during Reagan's visit, the news magazine *Der Spiegel* noted that these "siege tactics" were easy to accomplish because Kreuzberg's most unsettled neighborhood was like a "black triangle between the wall and Landwehr Canal," almost entirely cut off from the rest of the city already (June 22, 1987, p. 23).

The sense of isolation was so complete that in a series of books published on West Berlin's districts as part of the 1987 celebration of Berlin's 750th anniversary, which one might expect to be particularly complimentary, Kreuzberg was condemned as a "lifeless triangle" that was in danger of becoming a "peninsula" and "isolation ward for social problems and poverty" (Kaak, 1988, p.10). The occupation of the Lenné Triangle occurred in the context of this spate of recent violent confrontations pitting Kreuzberg against the rest of West Berlin. As itself a small triangle of neglected land, both marginalized and rendered nationally visible by its proximity to the Wall, the Lenné plot took on many of the features of Kreuzberg, in miniature. As Kreuzberg was Berlin's Berlin, Lenné became, so to speak, Kreuzberg's Kreuzberg. These themes of isolation and closure, the urban utopianism of rubble greenery and anarchic violence would all come to play in the conflict over the Lenné Triangle in the summer of 1988.

---

1 Here and throughout, translations from the German are my own.

THE LENNÉ TRIANGLE INCIDENT: THE POLITICS OF OCCUPATION IN NO MAN'S LAND

In February 1988 the Mayor of West Berlin, Eberhard Diepgen, met with the President of the German Democratic Republic, Erich Honecker, to discuss several issues of cross-border interest. The central component of the resulting agreement was the exchange of a number of small plots of land along the Berlin Wall, completing a process of regularizing the border and eliminating exclaves that had begun in 1971 (Barclay, 2012, p. 117). The key piece of land in the transaction was the four-hectare plot right off the Potsdamer Platz known as the Lenné Triangle. Although the land had always belonged to the GDR, it remained on the western side when the Wall was first constructed in 1961. The awkward angle at which it jutted into West Berlin would have made the construction of a wall around it difficult and costly. Though long neglected, the plot was considered of particular value because of its prime location. The West Berlin government was considering plans to use it in the construction of a major highway connecting the Reichstag in the north and the Landwehr Canal in the south, and traversing the length of the Tiergarten (*Der Spiegel*, March 28, 1988, p. 89). The land would formally change hands on July 1, 1988. In the intervening months, this now highly publicized and visible plot of land remained outside of the reach of the West Berlin police (*Frankfurter Allgemeine Zeitung*, June 2, 1988, p. 17).

Though this bit of land adjacent to Potsdamer Platz had a long and eventful history, by 1988 the Lenné Triangle was an empty space. Like other exclaves along the border, its existence as a no man's land was the product of the fact that the Wall itself had never perfectly followed the sector border. Patrolled by GDR border guards, it had remained off-limits to West Berliners and inaccessible to East Berliners. Mostly, the land was just left alone, though once in 1986 East Berlin border guards had crossed the wall to arrest a Kreuzberger for defacing the wall (Noé, 2008). So, over decades of disuse, the little plot had become one of those abandoned plots grown wild that so characterized Kreuzberg's environment as a wall neighborhood. By 1988, environmentalists in West Berlin claimed that the Triangle contained 161 different varieties of plants, including 12 listed as endangered on the "Red List" put out by the West Berlin Senate Department for Urban Development (*Frankfurter Allgemeine Zeitung*, June 16, 1988, p. 9). It was an example of what one politician of the time called "trümmergrün," or "rubble greenery": abandoned un-built land that became a kind of de facto urban green space, valued by some urban environmentalists (Strom, 2001, p. 197). The plan to clear the plot for the construction of a highway was seen by some West Berliners, and Kreuzbergers in particular, as an example of environmentally and socially insensitive policy typical of West German politics, endemic of a system that left no room for the unencumbered growth of either human communities or wildflowers.

After the planned exchange had been announced, the GDR removed the fence that had surrounded the patch of land. Wondering why they had not waited until the official exchange was completed in July, the West Berlin Interior Senator Wilhelm Kewenig joked that there must have been a dire shortage of fences in the

GDR (*Der Spiegel*, July 4, 1988, p. 75). This action allowed West Berliners new ease of access to the land before the actual exchange of territory occurred: it still belonged to the GDR, but was no longer off limits. It was out of reach of West Berlin police at the discretion of East German border guards. So, with the stated goal of preventing the triangle's use for the highway project, which would both endanger the wild flowers and butterflies in the small plot and also threaten the green space of the Tiergarten as a whole, a collection of Kreuzberg activists decided to turn the weedy plot into the location of a new squat.

The occupation of the Lenné Triangle began on May 25, 1988. The occupiers reached an agreement with the East Berlin border guards that allowed them to set up tents and later temporary huts constructed from wood and plastic tarpaulins around the perimeter of the triangle (*Die Tageszeitung*, May 30, 988, p. 17).[2] Over the next week, they planted potatoes and vegetables in the areas that surrounded the "wild green" in the center of the triangle (*taz*, May 31, 1988, p. 17). A flyer recruited like-minded West Berliners to help secure the land: "If you are in the mood for summer, life, laughter, and argument, then kindly come join us, so that we can secure this place" (*Frankfurter Allgemeine Zeitung*, June 2, 1988, p. 17). By early June, the occupiers had truly settled in and were making strides toward self-sufficiency: two goats and six hens had joined them and their vegetable gardens in this green idyll in the shadow of the Berlin Wall (*taz*, June 8, 1988, p. 24).

The most immediate advertised goal of the squat was to protest the planned highway construction, but the powerful symbolism of the occupation of this highly visible plot of land next to the Wall allowed a certain flexibility and capaciousness in the protest's significance. According to one flyer, "The occupation is directed not only against the destruction of green space for a highway, but against the system that holds such piggishness to be necessary." Another protester proclaimed, "We are not occupying the Triangle symbolically; rather, we will stay here and set it up as our living space [*Lebensraum*]" (*taz*, May 30, 1988, p. 17). Although the seriousness of the occupiers' commitment to ecological principles was called into question in news coverage, the squatters received a number of powerful supporters, including environmentalist groups.[3] One of the group's leaders and a representative of the Alternative List, Stephan Noé, later reported that the group's motives had been

---

[2] *Die Tageszeitung* is here after referred to as the *taz* in all notes, as it is familiarly known in Berlin. Since the left–alternative paper dedicated daily space to the Lenné protest, it provides the most detailed accounts of the day-to-day happenings in the Triangle.

[3] Senator of the Interior Wilhelm Kewenig (CDU), who had a history of taking a hardline with Kreuzberg squatters, claimed that the occupiers were merely criminals who had jumped the wall to hide their identities from police. "You don't do that," he was reported to say, "if you are just a conservationist" (*taz*, July 2, 1988, p. 25). However, in addition to political groups like the Alternative List (AL), the occupiers received support from the Citizens' Initiative West (*Bürgerinitiative West-Tagente*), organized specifically to resist the construction of the highway meant to cut through the triangle, and the Berlin Section of the Association for the Environment and Conservation (*taz*, May 30, 1988, p. 17).

"a mix of ecological demands and conscious provocation" (quoted in Pragel and Stratenschulte, 1999, p. 157).

Throughout the month of June, the occupiers worked to create an autonomous space representative of Kreuzberg's culture of protest. One of the first actions they took was to rename the plot the Norbert-Kubat-Triangle, after a demonstrator who had committed suicide in prison the previous year. Their numbers grew rapidly, though they also fluctuated with the weather, reaching over 100 by June 12. Playing up the theme of the new settlement as a miniature city, set apart from West Berlin, the alternative city daily *Die Tageszeitung* (*The Daily Newspaper*, hereafter "*taz*") referred somewhat tongue-in-cheek to the triangle's different generations of structures as the "Old Town" and "New Town" ("Altstadt" and "Neustadt"), the latter of which included both a "people's kitchen" ("Volxküche") and the "Rudi-Dutschke-House." The Triangle even had its own street of sorts, which the occupiers called the "Pesttangente," an allusion to the "Westtangente" highway the occupiers were protesting (*taz*, June 9, 1988, p. 24). Although satirical, these references acknowledged, and in some ways legitimated, the occupiers' attempt to create an autonomous space for political expression. They also played on the theme of the Lenné Triangle as a microcosm of West Berlin's own peculiar spatiality as an exclave.

As the occupiers sought to delineate the space as their own, the West Berlin police worked to define and separate the Triangle as well in their effort to control and contain the protesters. On May 30, only five days after the occupation began, the police cordoned off the Triangle with iron rods and red and white construction tape. The *taz* referred to the plastic cordons as a "symbolic 'ersatz-wall'" (*taz*, May 31, 1988, p. 17). The number of entrances into the plot was reduced to two, both closely guarded by police (*taz*, June 9, 1988, pp. 2, 17). The official reason for the cordoning was the possible presence of unexploded World War II ordinance underground, reinforcing the perception of the no-man's-land along the wall as a dangerous space with an ever-present potential for violence (*Der Spiegel*, June 27, 1988, p. 60; Nawrocki, 1988, p. 7). By mid June several papers referred to the plot as a "Wehrdorf," or fortified village, with its own watchtower, ditches, and protective walls in eerie mimicry in miniature of the barricaded status of West Berlin itself (*Frankfurter Allgemeine Zeitung*, June 13, 1988, p. 12).

The promptness of police action against the squat and their eagerness to control and demarcate the space with a fence betrayed anxiety about the plot's ambiguous jurisdiction, which itself reflected the instability of Berlin's supposedly impermeable internal border. Reflecting this anxiety, West Berlin's Interior Senator Kewenig expressed an especial indignation that the Triangle was a "territory that virtually lays outside of any jurisdiction" that could be occupied and fortified outside the city's control, but with materials taken from West Berlin (Nawrocki, 1988, p. 7). While the plot's jurisdictional ambiguity unsettled officials, it provided the occupiers with space for protest. "The state can assert itself everywhere," one occupier said in an interview, "we want to have a piece of land where we can assert ourselves too" (Tillack, 1988, p. 3). That the space they found was technically property of the East German state was one of the incident's ironies; there would have been no space and far less tolerance for their actions on the other side of the Wall.

Notably, then, the fence around the Triangle became the principal bone of contention between occupiers and the police. A week after putting up the initial cordons, the police replaced the red and white construction tape with a sturdier and more permanent fence (*taz*, June 8, 1988, p. 24). The occupiers demanded that the fence be dismantled as the only condition for their engagement in peaceful negotiations with an arbitrator. Their request was denied, ostensibly because of their prior violence (Plarre, 1988, p. 3). Attempts to tamper with or dismantle the fence to transport food or building materials were met by police resistance. Early on, the *taz* characterized these conflicts as an unequal battle between club-wielding police and the garden gnomes of the Lenné settlers, who simply wanted a peaceful space for their garden allotments (Witte, 1988, p. 24). Over the next weeks the conflicts escalated. The police expressed frustration that the British *Schutzmacht* stood by as the occupiers transformed the land into a "fortress" (*Festung*) (*taz*, June 16, 1988, p. 18). They used water cannons and tear gas to try to drive the occupiers off the land while the latter threw stones, scraps of metal and Molotov cocktails at police, culminating in heavy, several-hour-long conflicts on June 21–22 (*taz*, June 21, 1988, p. 2; June 22, 1988, pp. 1–2).

Both the police and the CDU-led city government came under heavy criticism for encouraging escalation rather than discussion with the protesters (*Frankfurter Allgemeine Zeitung*, June 22, 1988, p. 2; *taz*, June 21, 1988, p. 2). The police also used unconventional methods to try to drive the occupiers from the Lenné Triangle, since their actions were limited by the fact that the East Berlin border guards complained every time they strayed into East German territory. For example, starting early in the morning of June 8, police were reported to have tried to drive out the Triangle's occupiers by preventing them from sleeping with loud music (*taz*, June 9, 1988, p. 2). This, recalling the early realization that loud speakers were one of the few easy and reliable ways to bridge the Wall itself, was one of the many ways in which the Lenné incident turned the apparatus of the border inward. Fences, walls, and loud speakers were directed by West Germans against other West Germans, echoing the way in which the Wall itself had turned Germans against Germans. But while the earlier use of loud speakers in an attempt to convert East Germans implied intentionally gentle methods in the context of the potential for Cold War violence, now it highlighted the way in which the West Berlin police were treating the protesters themselves as the enemy, seeking to distance them from the rest of the city.

As the actions of police and squatters engaged the symbolism of the Lenné Triangle's unique geography, press coverage of the protest routinely played with the tension between the geopolitical significance of the squat's location and the localism of the protesters' aims. The national press covered the protest with a mixture of bemusement and anxious annoyance. An article that ran in the national news magazine *Die Zeit* a few weeks into the occupation essentially conflated the "colorful" protesters with the Wall's colorful graffiti:

> *The wall at Potsdamer Platz is as colorfully painted as ever. At its feet sit about a dozen of the Lenné-Triangle occupiers, mostly masked, and listening on a*

> *cassette recorder to the Berlin cabaret artist Martin Buchholz: ... Reagan asked Gorbachev: 'Why is the train stopping so long?' Answer: 'They just traded the locomotive for vodka.' Above on the top of the wall GDR border guards lean on their elbows; one looks through a telescope, the other considers the colorful activity (bunte Treiben) in no-man's land with amusement. (Nawrocki, 1988, p. 7)*

Depicting the occupiers as part of the Wall's bright decoration seems to defuse the power of their protest, and yet, as we know, graffiti on the Berlin Wall was serious business. As in the Martin Buchholz quip, levity and seriousness were impossible to disentangle in Cold War Berlin, where joking was serious business and ordinary life defied reason.

An article about the Triangle published in *Der Spiegel* played up the parallels between this small-scale event and other historical Berlin crises. The article opens describing an ominous scene: "Military helicopters rattle over the wall, tear gas grenades explode." On one side march "the East Berlin border guards in Russian gas masks" and on the other are the West Berlin police with their helmet and clubs: "So the situation escalates in the old familiar way." The title of the article itself, "Checkpoint Norbie," plays with reference to Checkpoint Charlie. ("Norbie" was in reference to Norbert Kubat.) But, the article goes on, in this case "what looks like a proper Berlin Crisis, is only a local tussle" (June 27, 1988, p. 60). Whether trivializing the protest or treating it as an ominous echo of serious crises, one thing that the media coverage of the Lenné incident never did was to take seriously the possible geopolitical implications of the protest itself: that the protesters needed a free space within West Berlin from which they could critique it and suggest a system of values other than those embodied by the highway project.

The transfer of the Triangle to the West Berlin government proceeded on schedule at midnight on July 1, 1988. Amidst widespread fear that the subsequent clearance could only be accomplished violently, it succeeded quickly and without a single injury on either side. Around five o'clock in the morning, just hours after the transfer, between 800 and 900 police surrounded the triangle, pressing the occupiers back into the two meter strip of East German land immediately in front of the Wall (*Frankfurter Allgemeine Zeitung*, July 2, 1988, p. 2; *taz*, July 2, 1988, p. 1). Once cornered, most of the occupiers simply climbed the Wall and then jumped over to the East Berlin side in order to escape the police. Others managed to escape through an opening in the fence right next to the Wall, and one was arrested for previous aggression against the police (*Frankfurter Allgemeine Zeitung*, July 2, 1988, p. 2). In the East, the jumpers were received by waiting flatbed trucks. According to reports, they were fed breakfast, asked to use one of the official checkpoints on their next visit to East Berlin, and sent back in small batches at various border crossings. The party organ of the GDR's ruling party, *Neues Deutschland*, which had previously commented on the civil war-like conditions in the West as clouds of tear gas wafted over the Wall near the triangle, reported merely that "the relevant authorities of the GDR were helpful in returning them back to Berlin (West)" (July 2, 1988, p. 7). By the very next day, bulldozers had removed almost all evidence of the six-week occupation. Some of the occupiers hung around, unsure of what to do next (Schmemann, 1988).

The 180-odd Wall jumpers were aware of the symbolic import of their action and proud of their ability to invert the normal machinery of images associated with the German–German border. Tourists had shown up to observe this authentic demonstration of West Berlin's counterculture, giving the occupiers an awareness of possible national and global audiences. One was quoted as saying, "When the Americans see the business about the wall jumping on their televisions today, then they can forget all that wall shit. It will turn everything on its head" (Tillack, 1988, p. 3). This occupier clearly understood that the power that the Wall invested in his actions was that of a megaphone: broadcasting his actions around the world. As Stephan Noé later recalled, the events at Lenné inevitably had a wider echo because of the presence of tourists who usually came to observe events on the other side of the Wall, but now directed some of their sympathy towards the occupiers on the nearside of the Wall (Noé, 2008). But in addition to providing a wider audience for their actions, Berlin's exceptional status actually served instead to empty their political protest of its specific meaning. Rather than inverting everything they thought their American audience might know about the Berlin Wall, the *New York Times* referred to the incident as "just a Berlin kind of day" (Schmemann, 1988). The national German press also noted the link between the Lenné incident and the peculiar histories of both Kreuzberg and Berlin (*Frankfurter Allgemeine Zeitung*, June 16, 1988, p. 9).[4]

Although the Lenné Triangle was particularly politicized because of its importance and visibility, it was not the only refuge in the shadow of the Wall. Similar concerns were raised over another of the plots exchanged in 1988, the transfer of which allowed the reopening of a canal bridge. Residents protested that the land along the canal would be destroyed by the inevitable increase in traffic. Graffiti scrawled on the side of the bridge proclaimed, "If no Wall then Auto Power—The Wall must remain" (*taz*, July 19, 1988, p. 15). Outside of the few organized efforts of environmentalists to lay claim to the wild environment on the border, Kreuzbergers quietly planted and cultivated gardens and playgrounds along the Wall (Nowakowski, 1988, p. 24). Spaces like these caused one Kreuzberger to lament how "the wall … [wa]s the only peaceful place in Berlin" (*taz*, August 15, 1988, p. 14). Without Lenné's highly significant central location, however, these other conflicts never came to have the same symbolic weight and wider visibility.

Marginalized groups, from Turkish immigrants to political radicals, to those searching for an alternative environmentally-friendly lifestyle, identified with the environment of the Wall as representative of a border space that allowed behavior and types of social organization for which there was no room in mainstream culture. This was symbolized in part by the wild, weeded environment that sprung up in the deserted and marginalized landscapes that skirted the edge of the wall. Although in the center of the city, the environment immediately adjacent to the Wall was unusable, and so created a peripheral and free space. In the Lenné incident, as in

4 On the Lenné Triangle's longer history, see McGee, 2002, pp. 198–204.

these other cases, however, the Wall's larger geopolitical significance complicated this fight to preserve these spaces. It was difficult to argue for the desirability of preserving the no man's land, regardless of the ecological benefits. "What should we do?" asked environmental activist Peter Kruse, "we certainly can't go down to the Ku'damm and demonstrate in favor of wall and barbed wire" (*Der Spiegel*, March 28, 1988, p. 91).

CONCLUSION

The events of the Lenné occupation demonstrate the distinctive ways in which the surveillance and spectacle of the Wall and the Cold War border it marked implicated global and local politics in the urban landscape of divided Berlin. Whereas earlier accounts have used the Lenné incident as direct evidence of the absurdity of divided Berlin's geography, this chapter has instead examined the ways in which different groups at the time deployed that same rhetoric of absurdity within political and cultural conflicts in West Berlin in the late 1980s. For the Lenné protesters, the geopolitical significance of the site was important because the location of their squat associated the oppressive West Berlin *Betonpolitik* ("concrete politics") against which they were protesting (in the form of the planned highway) with the city's most notorious concrete structure, the Berlin Wall itself. In response, the West Berlin police themselves became the ones building walls and reinforcing boundaries, recreating the injustices of the Wall within West Berlin. On the other hand, the media coverage of the event outside Berlin—whether nationally or internationally—used the absurdity of Berlin's divided geography to defuse the potential political impetus of the protesters' actions, allowing them to ignore their actual goals in favor of playing on themes of potential Cold War violence. From both sides, the effect was to reinforce West Berlin's exceptional status, instead of undermining the logic of the city's division.

In addition to illuminating the character of divided Berlin on the eve of the Wall's fall, this incident also provides insight into the functioning of border spaces more generally. Local and global politics became implicated in this particular site through a tendency to see spatial patterns as reproducing across scale (Gaddis, 2002, p. 84). Kreuzberg mimicked West Berlin, and the Lenné Triangle mimicked Kreuzberg. The inclination to "wall off the Wall" was reflected in the attempt to recreate walls—both mental and physical—around each the district of Kreuzberg and the Lenné Triangle itself, each labeled as a "lifeless triangle" (Kaak, 1988, p. 10). This is a powerful demonstration of the effect that borders have on those who live in their shadow, and how borders seem to legitimate their own presence by recasting the spaces around them.

REFERENCES

Barclay, D.E. (2012). "A 'Complicated Contrivance': West Berlin Behind the Wall, 1971–1989." In M. Silberman, K.E. Till and J. Ward (eds) *Walls, Borders, Boundaries: Spatial and Cultural Practices in Europe*. New York: Berghahn Books. pp. 113–30.

Borneman, J. (1992). *Belonging in the Two Berlins: Kin, State, Nation*. Cambridge: Cambridge University Press.

Colomb, C. (2011). *Staging the New Berlin: Place Marketing and the Politics of Urban Reinvention Post-1989*. New York: Routledge.

Flemming, T. and Koch, H. (2005). *Die Berliner Mauer: Geschichte eines politischen Bauwerks*. Berlin: Bebra Verlag.

Frankfurter Allgemeine Zeitung (1988). "Berliner Polizisten kritisieren Kewenig." *Frankfurter Allgemeine Zeitung*, 22 June, p. 2.

Frankfurter Allgemeine Zeitung (1988). "Das Lenné-Dreieck: Nicht Nur ein Kuriosum der deutschen Teilung." *Frankfurter Allgemeine Zeitung*, 16 June, p. 9.

Frankfurter Allgemeine Zeitung (1988). "Der DDR erlaubt den Besetzern Spaziergänge." *Frankfurter Allgemeine Zeitung*, 2 June, p. 17.

Frankfurter Allgemeine Zeitung (1988). "Ein explosives Stuck Land." *Frankfurter Allgemeine Zeitung*, 13 June, p. 12.

Frankfurter Allgemeine Zeitung (1988). "Freundliche DDR Grenzpolizisten ziehen die Besetzer auf die Mauer." *Frankfurter Allgemeine Zeitung*, 2 July, p. 2.

Frankfurter Allgemeine Zeitung (1988). "Weiter Auseinandersetzungen am Lenné-Dreieck." *Frankfurter Allgemeine Zeitung*, 21 June, p. 6.

Gaddis, J.L. (2002). *The Landscape of History: How Historians Map the Past*. Oxford: Oxford University Press.

Haxthausen, C.W. and Suhr, H. (1990). "Introduction." In C.W. Haxthausen and H. Suhr (eds) *Berlin: Culture and Metropolis*. Minneapolis, MN: University of Minnesota Press, pp. xi–xxiii.

Kaak, H. (1988). *Geschichte der Berliner Verwaltungsbezirke 2: Kreuzberg*. Berlin: Colloquium Verlag.

Karapin, R. (2007). *Protest Politics in Germany: Movements on the Left and Right since the 1960s*. University Park, PA: Pennsylvania State University.

Ladd, B. (1997). *The Ghosts of Berlin: Confronting German History in the Urban Landscape*. Chicago, IL: University of Chicago Press.

Lang, B. (1998). *Mythos Kreuzberg: Ethnographie eines Stadtteils (1961–1995)*. Frankfurt am Main: Campus Verlag.

Large, D.C. (2000). *Berlin*. New York: Basic Books.

Mayer, S. (1996). "The Graffiti of the Berlin Wall: A Semiotic Approach." In E. Schürer, M. Keune and P. Jenkins (eds) *The Berlin Wall: Representations and Perspectives*. New York: Peter Lang Verlag, pp. 214–28.

McGee, M.R. (2002). *Berlin: A Visual and Historical Documentation from 1925 to the Present*. Woodstock, NY: The Overlook Press.

Nawrocki, J. (1988). "Lenné-Dreieck in Berlin: Ärgerliche Attraktion." *Die Zeit*, 17 June, p. 7.

Neues Deutschland (1988). "Lenné-Dreieck von Westberliner Polizei mit Gewalt geräumt." *Neues Deutschland*, 2 July, p. 7.

Noé, S. (2008). "Kalter Krieg bizarr: Über die Mauer in den Osten." *Spiegel Online*. Available at: http://einestages.spiegel.de/static/authoralbumbackground/752/ueber_die_mauer_in_den_osten.html (accessed 7 October 2013).

Nowakowski, G. (1988). "Gute Erde." *Die Tageszeitung*, 20 July, p. 24.

Plarre, P. (1988). "5 Wochen Kubat-Dreieck." *Die Tageszeitung*, 2 July, p. 3.

Pragel, P. and Stratenschulte, E.D. (1999). *Der Monolog der Lautsprecher und andere Geschichten aus dem geteilten Berlin mit einem Vorwort von Klaus Schütz*. Munich: Deutscher Taschenbuch Verlag.

Pugh, E. (2010). "Beyond the Berlin myth: The local, the global and the IBA 87." In P. Broadbend and S. Hake (eds) *Berlin Divided City, 1945–1989*. New York, NY: Berghahn Books, pp. 156–67.

Richie, A. (1998). *Faust's Metropolis: A History of Berlin*. New York: Carroll & Graf.

Scheffler, K. (1910). *Berlin: Ein Stadtschicksal*. Berlin-Westend: Erich Reiss Verlag.

Schmemann, S. (1988). "Turf Trade and a Tussle: Just a Berlin Kind of Day." *New York Times*, 2 July [online]. Available at: http://www.nytimes.com/1988/07/02/world/turf-trade-and-a-tussle-just-a-berlin-kind-of-day.html (accessed 7 October 2013).

Schneider, P. (1983). *The Wall Jumper*. Translated from German by L. Hafrey. New York: Pantheon Books.

Spiegel (1987). "Ordentlich scheppern." *Der Spiegel*, 22 June, pp. 23–4 [online]. Available at: http://www.spiegel.de/spiegel/print/d-13524765.html (accessed 25 January 2014).

Spiegel (1988). "Checkpoint Norbie." *Der Spiegel*, 27 June, pp. 60–61 [online]. Available at: http://www.spiegel.de/spiegel/print/d-13529827.html (accessed 25 January 2014).

Spiegel (1988). "Honecker 2X klingeln." *Der Spiegel*, 28 March, pp. 89–91 [online]. Available at: http://www.spiegel.de/spiegel/print/d-13528857.html (accessed 25 January 2014).

Spiegel (1988). "Rache kalt." *Der Spiegel*, 4 July, p. 75 [online]. Available at: http://www.spiegel.de/spiegel/print/d-13528096.html (accessed 25 January 2014).

Steinfeld, T. (1990). "Afterword: Writing about Berlin." In C.W. Haxthausen and H. Suhr (eds) *Berlin: Culture and Metropolis*. Minneapolis, MN: University of Minnesota Press, pp. 253–6.

Strom, E.A. (2001). *Building the New Berlin: The Politics of Urban Development in Germany's Capital City*. Lanham, MD: Lexington Books.

Tageszeitung (1988). "Die Mauer muß bleiben." *Die Tageszeitung*, 19 July, p. 15.

Tageszeitung (1988). "Dreieck unter Gas und Wasser." *Die Tageszeitung*, 21 June, p. 2.

Tageszeitung (1988). "Ein zufriedener, aber beleidigter Feldherr." *Die Tageszeitung*, 2 July, p. 25.

Tageszeitung (1988). "Massenflucht über die Mauer—in den Osten." *Die Tageszeitung*, 2 July, p. 1.

Tageszeitung (1988). "Mauer vergessen? Auf der Suche nach einem Standbild." *Die Tageszeitung*, 15 August, p. 14.

Tageszeitung (1988). "Pesttagente." *Die Tageszeitung*, 9 June, p. 24.

Tageszeitung (1988). "Plastikband als Mauerersatz." *Die Tageszeitung*, 31 May, p. 17.

Tageszeitung (1988). "Sorgenvolle Polizei." *Die Tageszeitung*, 16 June, p. 18.

Tageszeitung (1988). "Unter Wasser." *Die Tageszeitung*, 9 June, p. 17.

Tageszeitung (1988). "Vor der Mauer liegt der Strand." *Die Tageszeitung*, 30 May, p. 17.

Tageszeitung (1988). "Wenig Schlaf für Lenné-BesetzerInnen." *Die Tageszeitung*, 9 June, p. 2.

Tageszeitung (1988). "West-Berlins Polizei bringt DDR-Grenzer zum Heulen." *Die Tageszeitung*, 22 June, pp. 1–2.

Tageszeitung (1988). "Zwei Ziegen und Sechs Hühner." *Die Tageszeitung*, 8 June, p. 24.

Taylor, F. (2006). *The Berlin Wall: A World Divided, 1961–1989*. London: Bloomsbury.

Tillack, H.-M. (1988). "Mauersprung von West nach Ost." *Die Tageszeitung*, 2 July, p. 3.

Ward, J. (2011). *Post-Wall Berlin: Borders, Space and Identity*. New York: Palgrave Macmillan.

Webber, A. (2008). *Berlin in the Twentieth Century: A Cultural Topography*. Cambridge: Cambridge University Press.

Witte, J. (1988). "Landgewinnung: Wenn die Polizei auf die Zwerge losginge." *Die Tageszeitung*, 9 June, p. 24.

3

Remaking the Edges: Surveillance and Flows in Sub-Saharan Africa's New Suburbs

Garth Myers

INTRODUCTION

The Millennium Village occupies a large plot in the heart of Lusaka's colonial-era central administrative area, with its gate on the tree-lined Birdcage Walk, a short way from what the city's British planners called the Ridgeway, where the colonial administrators sat atop the city (Myers, 2003). The Millennium Village, however, was purpose-built with funding from the government of Libya to coincide with the 2000 meeting of the African Union. Libya's leader at that time, the late Muammar Gaddafi, had a house built for each member state's President, with the name of the country emblazoned on the front above its door.

Thirteen years after the Millennium, the Village had a very different character. It had become a gated corporate business park, dominated by Chinese engineering and development firms. The house originally intended for Tanzania's President now had become the offices of the Henan-Guoji Development Company (HGDC) in Zambia (Figure 3.1). As of 2013, HGDC had housing development projects underway in nine African countries, including two in Zambia. All came under the name of a "Guoji Dream Town," a highly controlled exclusive community in the suburbs of a given African city. Lusaka's version, called Silverest Gardens, included 380 home sites with eight different house models "in a tranquil place away from the city buzz" and outside of the legal boundary of Lusaka city (HGDC, 2012) (Figure 3.2).

The transformation of the Millennium Village and the creation of Silverest Gardens typify the topographies of surveillance and flows emerging across urban Africa. Urban frontiers are in a state of dramatic change across the continent. From Accra to Zanzibar, from Dakar to Durban, urban growth long ago jumped the borders of the city, but the last decade has seen a mushrooming of new suburbs, satellite towns, and gated communities, alongside expanding peri-urban informal settlements (de Boeck, 2012; Huchzermeyer, 2011; UN Habitat, 2010). This urban expansion is, indeed, fairly consistently bifurcated. On one hand, middle-

3.1 The Henan-Guoji Development Company headquarters, in Tanzania House, Millennium Village Business Park, Lusaka, June 2013. All images (3.1–3.7) courtesy of Garth Myers.

and upper-income housing areas like Silverest Gardens are appearing across the continent, seemingly out of step with a slower pace of economic growth and income expansion. This side of the urban footprint's expansion has gathered steam in the ongoing New Scramble for Africa, particularly with Chinese investment in new real estate ventures and urban infrastructure (Carmody, 2011; Carmody and Hampwaye, 2010). The rise of surveillance technologies, particularly in new gated communities, is hard to miss, but so are the flows and connections across the globe for both residents and investors in these properties (Murray, 2011 and 2013; Bénit-Gbaffou et al., 2012).

On the other hand, larger numbers of urban residents continue to reside in ever increasing marginalized informal settlements. In many cities, the growth of investment in elite and middle class real estate has established a sort of chain of displacement leading to the growth of slums, often still further from the centers of cities. In Nairobi and Dakar, for example, elite property development in inner-ring western suburbs, including both malls and luxury mid-to-high-rise condominiums as well as single-family fortresses, has pushed many working class or lower-middle class former residents into informal settlements in places like Kibera (in Nairobi) and Pikine (outside Dakar). Rising prices within informal areas then displace the more marginal populations in them into even more marginalized housing circumstances

3.2 Billboard promotion for Silverest Gardens, Lusaka, June 2013.

(Huchzermeyer, 2011; Diouck, 2013; Myers, 2015; Ngau, 2012; Bousquet, 2010). As of 2013, Pikine, for instance, had incredible overcrowding in areas with chronic, severe seasonal flooding—so that 30 or 40 people reside in a two-bedroom house with a half-meter of standing water on its ground-floor level in the rainy season (Fall, 2013). Spaces like these reinforce the underlying continuity of urban development patterns across Africa *now* with those of the colonial era, though, particularly in terms of the odd mix of lax controls with capricious, sporadic bursts of enforcement which characterize everyday life at the urban borders (Myers and Murray, 2006).

This chapter examines the new edge areas of Lusaka, based on fieldwork from 2013, as a broad example of the trajectory of urban expansion at the new urban frontiers of Africa. Even a cursory assessment of Lusaka in comparison with literature for cities like Dakar, Dar es Salaam, Johannesburg, Kinshasa, Accra, Lagos, and Nairobi, suggests that there may be some differences in how things are playing out across the continent. For example, Dakar's far-flung diaspora of traders associated with the Mouride Islamic brotherhood invest earnings from New York or Guangzhou in new apartment buildings back home; Accra or Lagos must contend with substantial legal cases over indigenous land claims in the expansion zone; and the post-apartheid era's complex interplays of race and class in the unicity governance of Johannesburg have little in common with Kinshasa's piratic and ungovernable growth patterns (Myers, 2011; Grant, 2009; Ardayfio-Schandorf et al., 2012; Akinyele, 2009; Trefon, 2009). Yet I argue that there are also themes that can be held in common in the region's cities. I emphasize four of these in the Lusaka case: (1) the significance of new sources for flows of foreign investment in urban frontier zones; (2) the bifurcated character of the expansion; (3) the rise of regimes of surveillance; and (4) the endurance of continuities with European colonialism in the ambivalence and contradictions of urban borders. First, though, I provide a brief introduction to Lusaka.

A GARDEN CITY FOR AFRICA AFTER A CENTURY

Lusaka is a city of 1.7 million people as of the latest national census, with most scholars and planners who study it estimating that its actual population is well over two million. It began as a railway watering-station settlement in 1905 on the site of a Lenje village whose local leader, Lusaaka, gave the town its name; it officially became a town in 1913, serving a very small community of white settler farmers in the area. In the 1930s, this small town was utterly transformed with the establishment of Lusaka as the capital of the colony of Northern Rhodesia. The capital was redesigned along the general ideas of a Garden City for Africa, and it was opened as the new capital by King George V to much fanfare in 1935 (Bradley, 1935). Working from a template suggested by the architect and town planner S.D. Adshead, local colonial officials—mainly engineer P.J. Bowling and assistant chief secretary Eric Dutton—fashioned and built a layout for a low-density, white-only town of large lots, wide, spacious parklands, tree-lined boulevards, and the prominent new central government area, the Ridgeway, on the town's highest rise (Myers, 2003, pp. 55–6).

While in Lusaka "the imported values of the colonial power were translated into the physical form of a city," in reality the urban project was something of a flop (Collins, 1977, p. 227). The city grew slowly, and it grew in completely unintended shapes and patterns. As of 1928, prior to the capital construction, the official population of whites and Africans was less than 2,000; by 1946 the population was still under 20,000, with fewer than 8,000 African residents in formal employment. Bowling and Dutton's plan had included only a Governor's Village as a model settlement and a village for the African servants of government officials; both were highly controlled, surveilled and under-serviced settlements of tiny, round (rondavel) huts. Dutton's aim in these areas from a design standpoint was to preserve "what is best in the traditional plan of the African village" (Bradley, 1935, p. 47). Yet both areas were carefully situated; the Governor's Village was wedged in next to the military barracks, and the personal servants' compound was placed across the railroad tracks and on the down-slope from the Ridgeway itself. And, crucially, the planned garden city provided no other lands for African residential construction, when the city's African population was already in 1935 far larger than that of its white settler community (Myers, 2003, p. 66).

The first real burst of growth occurred after the Second World War, with the population, almost all of it African, reaching nearly 200,000 by independence in 1964 (Myers, 2003; Hansen, 1997). Thereafter the growth rate became dramatically higher, with more than half a million city residents by 1980, just under a million 10 years later and 1.7 million by 2010. The spatial frame of the capital area remained largely as planned, as did the architecture of government buildings. Planned, formerly white-only suburban townships of spacious homes on sizeable lots persisted outside the capital zone to its east and north. Almost nothing remains now, however, of the two planned African areas, except for a tiny segment of round huts that serve, ironically, as a government-sponsored "cultural village" for tourists (Myers, 2003, p. 67). The rest of Lusaka's people—approximately two thirds—came to reside in informal neighborhoods which the colonial regime had designated as "unauthorized areas" (Myers, 2006, p. 293). The colonial roots of Lusaka's deeply bifurcated spatial form spilled over into the contemporary setting for a variety of reasons, some unique to Lusaka, and others common across post-colonial Africa (Figure 3.3).

Even after the post-independence burst of population growth, Lusaka, like many African cities, largely failed to grow economically (Potts, 2004). Most of Zambia's economy has revolved around copper mining since Northern Rhodesia was created by Cecil Rhodes' mining conglomerate, the British South Africa Company, and most of that copper mining takes place on Zambia's Copperbelt. Lusaka has had a comparably small industrial base of light manufacturing, cement, and agro-processing; it is essentially a government town—the Ridgeway government area is a more impressive center than the drab CBD, as it was planned to be. Most of Lusaka's residents reside on land which had been designated as belonging to white-owned commercial farms or industrial sites outside of the restricted colonial capital site. Since most of the white settlers' economic ventures were modest, they rented land to Africans for the construction of homes. Technically, since the only Africans who were allowed to live in such areas were the employees of the white settlers (who were authorized to house their employees on their so-called compounds) such neighborhoods were unauthorized at their origins. This explains the unique designation of Lusaka's informal settlements as "compounds," and it explains why so many of these settlements, to this day, bear the first names, surnames, nicknames or family members' names of those white settlers.[1]

Because these unauthorized compound areas were outside the legal boundaries of the colonial government's official planning area for the capital in the 1930s, by the 1950s they were given the additional moniker of "peri-urban" areas, even when some were, in fact, just next to downtown Lusaka. Colonial and post-colonial government attempts to replace the unauthorized, informal, peri-urban

[1] Thus we see the major peri-urban compounds (neighborhoods) of John Howard, John Laing, George, Marapodi (for Italian cement contractor G.B. Marapodi), Mandevu ("Beards," Marapodi's nickname), Villa Elisabetta (for Marapodi's daughter, Elizabeth), or Misisi (The Mrs). Another, Ng'ombe (cows), takes its name from the area's role as a cow pasture for a local white farmer.

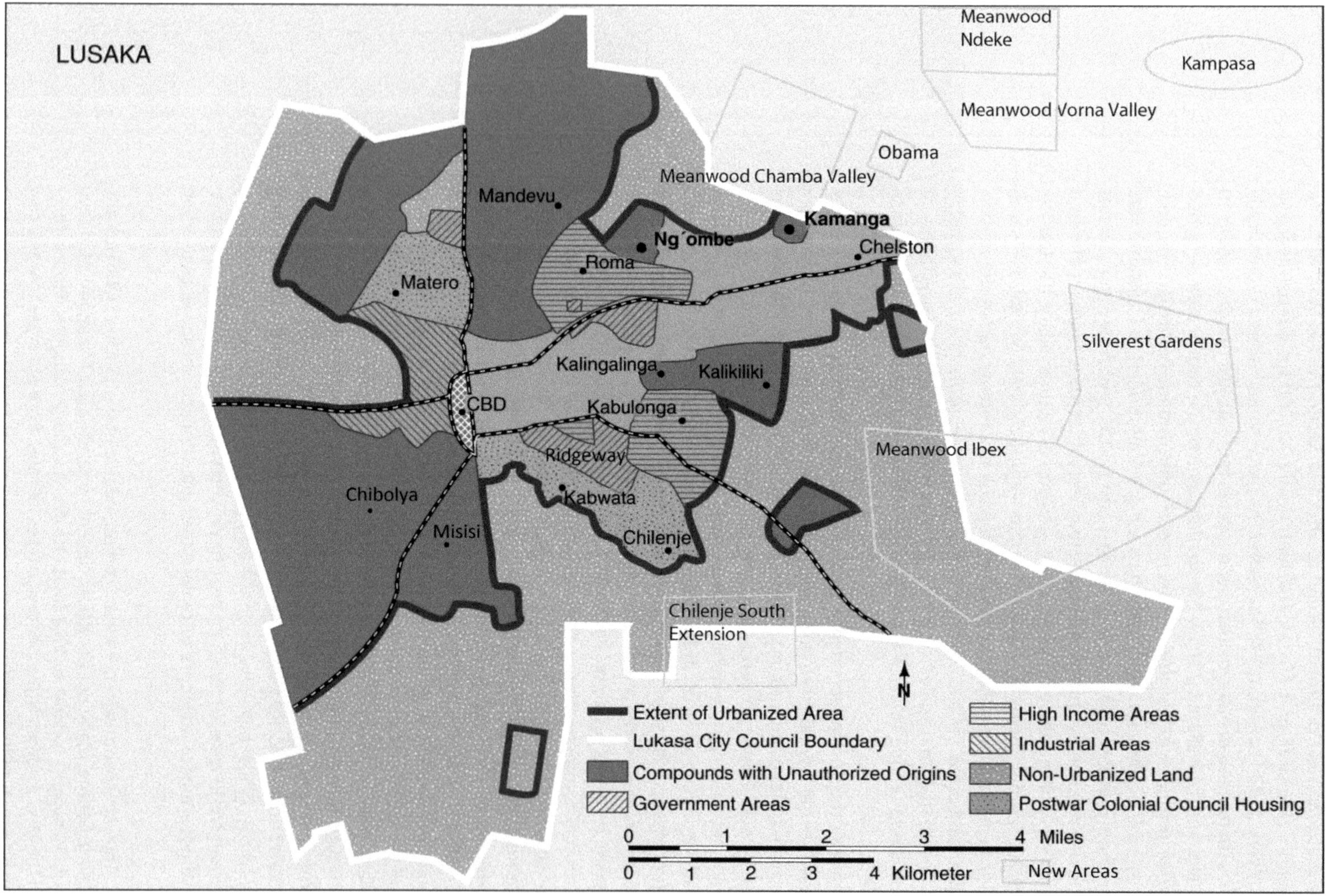

3.3 Map of Lusaka.

compounds with planned neighborhoods of government housing repeatedly fell far short in their goals, with new informal housing areas continuing to grow far more rapidly than the formal areas, and the subsidence of planned areas into a state of informality (Myers, 2005 and 2011). Lusaka essentially subsists in a post-colonial hangover state as a largely "unauthorized" city (Myers, 2006). The green veneer of the Garden City mystique still masks what it has always masked: a dusty, inelegant and largely poor city made up of a checkerboard of large, low-density planned elite townships (massive footprints, low populations) and high-density informal compound areas (*komboni* in Chinyanja, the city's *lingua franca*). Some of the terminology, like compound/*komboni*, may be particular to Lusaka, but a similar bifurcation of the city into broadly formal and informal housing zones which increasingly blur into one another around the poorly managed—indeed, unmanaged—urban edges afflicts the spatial form of most cities in Sub-Saharan Africa. That affliction seems to be worsening, in new ways, in the twenty-first century.

LUSAKA IN THE NEW SCRAMBLE FOR AFRICA

This new century has begun with an extraordinary explosion of investment and influence from China in Africa in general, and Zambia in particular. China has become Zambia's leading trading partner, donor and investor. Much of the initial impetus for Chinese interest in Zambia came from its strategic drive for mineral resources, and particularly copper (Carmody, 2011). But Chinese investments have broadened in scope to include agriculture and land development, alongside Chinese engineering firms which have placed successful bids for road and railroad construction. Significantly, firms like Henan Guoji Development Company have emerged in the last few years with residential real estate projects in the suburbs of Lusaka and the Copperbelt. Zambia's political and social stability, anchored and manifested in Lusaka, has combined with the emerging sense of other investors seeking to compete with the Chinese, to fuel expansion in foreign, local or transnational Zambian elite stakeholders in suburban Lusaka land and housing for both speculation and residence.

Indeed, driving or walking all around Lusaka in 2013, the first thing one would have seen everywhere would have been middle- and upper-class housing being built. Many of the ongoing projects were examples of urban infilling, as the city's plentiful array of open spaces from the colonial era was sold off and built up. But far more developments were occurring around the city's edges, mostly surrounding but sometimes on top of the colonial-era's peri-urban compounds. In Chilenje South, Woodlands South Extension, Chalala, Nyumba Yanga, and other areas along or across the city boundary, middle class and elite property development exploded in the last five years. Lusaka South Extension had a private planned development for 450 homes going in as of June 2013, with Chinese contractors. Far out Leopards Hill Road southeast of Lusaka, the Muka Munya development was a gated property for the top elites at the former picnic grounds of colonial elites around the natural

3.4 Billboard for Meanwood Vorna Valley and Ndeke, Lusaka, June 2013.

springs called Monkey Pools. There were 80 surveyed plots for sale in two areas separated by a wildlife park and nature reserve that also had a clubhouse with a pool and social hall (Nchito, 2013). Leopards Hill Road by 2013 had a set of other places catering to whites, and often white South Africans, such as Sugarbush, where even the building for the inn/restaurant/boutique was styled after a Cape Dutch (South African) winery; an upscale mini-mall housed an Italian/Zambian ice-cream shop, just down from the American International School.

Scattered around the edges of Lusaka to the northeast, east, and southeast, three distinct suburban projects were underway by mid-2013 under the name, Meanwood; this property development firm grew out of Galounia Farm, owned by the Galoun family. The Galouns were one of Lusaka's largest white landholders from the 1930s onward, but unlike many of their neighbors, they had not "farmed" their land out to unauthorized informal settlements. As the city grew around the farm's three main segments, the value of the land for residential construction skyrocketed. There will eventually be 5,500 homes in the Meanwood development in Ndeke-Vorna Valley of Chongwe District and another 5,500 in Chamba Valley just within Lusaka District (Figure 3.4); although many of these are more middle class than elite homes, and their residential character as of June 2013 was decidedly mixed (see below), the standard of housing still is well above that of the average Lusaka compound residents (Figure 3.5) (Myers and Subulwa, 2014).

3.5 Middle-class housing under construction, Meanwood Vorna Valley, Lusaka, June 2013.

Even still, the contrast between Meanwood Ndeke or Chamba Valley and ultra-elite Meanwood Ibex, the third Galounia Farm section to be developed into residential estates, is dramatic. This gated Meanwood lies to the northeast of the elite formerly white-only township of Kabulonga. Plot sizes are relatively small compared to the dramatically better houses, which have fixed building requirements from tile roofs to private security company contracts with Securicor or Armorcor. There are also some homes with electrified fences, but the walls or fences are shorter than those of nearby Kabulonga: residents and guests are meant to see these homes and their more elaborate landscaping, since only the privileged and their servants are ever expected to be in Meanwood Ibex (Figure 3.6). The back gate is essentially guarded by a gated community developed by the office of the President for Zambia's security service, and by Silverest Gardens (Myers and Subulwa, 2014). The front gate will eventually be a large wall separating Meanwood Ibex from the informal settlement just to its west, Kalikiliki.

A variety of elite (foreign and local) firms and agents have been at work in the creation of these new edge neighborhoods, but within Lusaka, much of the conversation among academics, officials and ordinary citizens alike in 2006–13 concerned the Chinese presence. China's role in the Zambian economy and political system became a key factor in the last two Presidential elections in Zambia. The losing candidate in the first of these and winner in the second, the late Michael

3.6 Elite housing nearing completion, Meanwood Ibex, Lusaka, June 2013.

Sata, capitalized on anti-Chinese sentiments in his campaigns, but then very soon after arriving at State House, Sata reversed course and strengthened relations with China. There was no doubt that he did so because Zambia's economy had become so thoroughly interdependent with that of China (Chilufya, 2013). But the ordinary conversations amongst Lusaka's African population were more significant for this chapter, because they typically did not revolve around high-level finance, geopolitics, or national identity. Instead, Zambians in Lusaka at barber shops and street vendor stalls in 2013 talked more frequently about Chinese families they observed while shopping, Chinese-Zambian children appearing in the back alleys of the compounds, and the subtle Chinese imprint in retail and in neighborhood design and housing styles, particularly in neighborhoods like Meanwood Ibex and, of course, Silverest Gardens.

TWO DIFFERENT TRAJECTORIES OF EXPANSION

If the intentionally visible manifestations of the New Scramble for Lusaka consist of developments like the three Meanwoods, more hidden trends become evident the more one looks. First of all, the imagined or intended Meanwood was a far cry from the actually existing Meanwood as of 2013. In Chamba Valley and

3.7 Signposts for informal employment, Meanwood Ndeke, Lusaka, June 2013.

Ndeke especially, many homes were barely half-finished in June 2013, with most being squatted by the workers building them. An entirely separate economy of construction gave informal life to the half-completed and ultimately only semi-planned neighborhoods. Meanwood Ndeke did have one visible open area planned for public space, but its only "market" consisted of some women selling vegetables and roasted corn at the entrance. Handpainted signs were abundant in the neighborhood for hairdressers, tailors, shoe repair, auto repair, "cooka" repair, "barba" shop, lawn mowing, eggs, borehole drilling, "tale fiter," and "panter" (Figure 3.7). There were many clotheslines full of drying clothes, and none of the roads were paved. None of the roads had any avenue trees in what was projected as the newest superficial extension of the Garden City idea (Myers and Subulwa, 2014).

At the other end of the income scale, informal settlements in peri-urban Lusaka continue to grow along with these temporarily informalized formal planned suburbs. As informal settlements like Ng'ombe (just south of Chamba Valley) are bought up or seized and then redeveloped into formal areas, Ng'ombe's poor find houses in the next of many "overspill" areas of unauthorized housing further to the east, in Kamanga or, since its emergence in 2009, in the Obama area north of Chelston. Indeed, many resident squatters in the new Meanwood plots in 2013 were former residents of the informal settlements that these new developments displaced, such as the former Ndeke village. And far from the new elite areas, somewhat hidden in plain view, one would find expansion of the lowest-income informal settlements, such those just south of Lusaka's downtown, in Misisi or Chibolya (Myers and Subulwa, 2014).

Lusaka's bifurcation into the Kabulonga/Meanwood sorts of areas and the Misisi/Chibolya types has long extended into its biogeography and landscape architecture (Pullan, 1986). In the new gated communities and planned extensions, the dominant tree and flowering plant species are ornamental exotics, mirroring the plantings of the formerly white-only townships. In the older unplanned settlements or semi-planned lower-income areas, such trees as one finds are nearly entirely fruit trees—even the hedge plants are typically fruit-bearing—along with food crop plants. In the poorest and newest squatter settlements, there are no trees, and barely any plants at all. Ninety-six percent of all properties in the elite suburban area of Kalundu in 2013 had a cement wall around the property, and most had either trees or hedge plants outside that wall. Only 22 percent of the properties in the older peri-urban compound of Kalingalinga had cement walls to mark the property boundary, and just 11 percent in the more recent slum of Misisi had such cement walls. While 47 percent of Kalingalinga residents at least had a hedge to mark the edges of their properties, scarcely 15 percent of Misisi properties did; two-thirds had no marking at all for the property borders (Myers and Subulwa, 2014).

These unequal spaces of Lusaka can also clash. The most notable example from 2013 occurred in a peri-urban area called Kampasa, just east from Meanwood Ndeke. Kampasa was an informal rural area straddling the border of Galounia Farm and agricultural land belonging to the Zambia National Service (ZNS). The ZNS gave 50 hectares and the Galoun family gave 30 hectares to a Chinese company for a commercial vegetable farming project. Without warning, the ZNS defense forces arrived in Kampasa at four in the morning and began to demolish houses. When residents spontaneously protested this action, the ZNS opened fire, killing two of the demonstrators (Mulenga, 2013). The Kampasa compound consisted of people who had done day labor for years for either the ZNS or for Galounia Farm (Anonymous interview, 2013). The Kampasa shooting provided a graphic and jarring example of the ever increasing social and spatial stratification in Lusaka, mainly between the people at the bottom in the compounds and those in either the middle or upper classes, and most dramatically around the urban edges. There is less of a clear spatial separation between the top and the middle classes, where both are "locking themselves away [in new developments]. They are kind of like prisons" (Kapungwe, 2013). For those trapped in the desolate poverty of Kampasa or Misisi, too, we find the built environment of a carceral city.

"SECURITY IS VERY IMPORTANT TO US"

The carceral city of urban theory is one obsessed with security, and the new planned suburbs in Lusaka manifest this, even in their strategies for promotion. Henan-Guoji Development Company's plan for Silverest Gardens literally occupied the central square of the upscale, CCTV-equipped and security-enhanced Manda Hill Shopping Mall on Great East Road for most of 2013. The company's three-dimensional plaster model of the community, equipped with a light board, was manufactured in China, showed an idealized portrait of what will be an "upgrade in habitation and revolution in LIVING," according to the brochure (HGDC, 2012). The brochure lays out the details of the development and highlights its capacity for changing Lusaka and transforming the city's individual residents. It touts the excellent location, six kilometers east of the Airport Roundabout, and just under two kilometers from the Great East Road out of the city, of course, but the tag line on the brochure's front offers this: "The Community changes the city." What the company means is: "using community development as a basic factor in gradually affecting city planning and development." They envision a process of "transforming the city with scattered residential layout into a modern city with perfect functionally, rationally planned and comfortable living environment." The eight model house types range from two to five bedrooms and from 84 to 220 square meters. Each style is promoted in its own way. While one two-bedroom home's "bright design is for the health of the inhabitant," the other "can make you enjoy the endless nature." Both "privacy" and "the taste of the master" or "the owner's noble status" are common promotional themes (all quotations from HGDC, 2012).

Life in Silverest Gardens is imagined as fully self-contained. It will eventually have its own shopping mall, kindergartens, police post, gym, social clubs, and services for landscaping, outside yard cleaning, waste collection and home maintenance. Its developers call it an excellent bargain and investment opportunity that is "delivering a new approach to urbanization and town design based on integrity and amicability." The brochure's numerous grammatical errors are matched by the company's extremely odd promotional video that is required watching for potential investors (Cornhill, 2013), replete with the imagined members of the new community marching in lockstep like members of the People's Army.

Fully 220 of the 380 homes had sold within the first six months of the promotion. All but one of the lowest-price models were sold by June 2013. Investors (most expatriates or transnational Zambians) snapped up the affordable ones, some buying 10 or 12 at a time. The highest priced home sites also sold out, again with an international cast of investors (Cornhill, 2013). Part of the scheme's great success with investors came from its security plan; there will be 24-hour all-over CCTV. HGDC sales agent Loreto Cornhill (2013) stressed that "security is very important to us." This is beyond the CCTV security, though; it is inclusive of the sense of security which comes with a clear, enforced spatial order. To wit, the company was waiting to do the roads and landscaping until all of the houses are sold. The plan was to be carried out by June 2014 all at once, to provide uniformity. There will be a shopping area in the center of the community so that one might never have to leave. HGDC will not let people change the exterior of homes, and they will manage the initial landscaping and take care of maintenance of the yards.

The idea of Silverest Gardens, then, is to create a neighborhood community that will somehow remake urban society and the city by locking everyone else out and by clarifying what exactly every square meter of its urban space can and cannot be. That the boundary of Silverest Gardens is shared on one side with the expansive Meanwood Ibex development means that the city's expansion is effectively walled off for the entire extent of the wedge from the Great East Road to Leopards Hill Road.

NEO-COLONIALISM AND THE PERI-URBAN

For all that places like Silverest Gardens purportedly offer something new in Lusaka, much remains similar to the colonial era. Tait (1997, p. 162) noted that "perhaps more than in other former colonial cities, the contemporary process of urbanization in Lusaka and its administrative framework remains rooted in the structures underpinning its historical expansion." Rakodi (1986, p. 213) cited the "structures, attitudes and feelings" of colonial planning, and the "underlying ideologies of separate development" for the supposed Garden City as crucial, tangible aspects of the colonial legacy in the city. Despite the appearances of so many distinctions—widespread middle- and upper-class high-standard private sector housing development, considered a major step of progress for mainstream economists (Collier and Venables, 2014), might be seen as the largest distinction of all—one can make the case that continuities with colonialism in structures, attitudes and feelings, and in a reformulated ideology of separate development, still anchor the new trends.

First, Lusaka remains a city with graphic spatial injustices, and the new wealthy housing areas on its outskirts only exacerbate these. Colonialism's influx controls and white-only townships are officially gone, but they are replaced with walls. At the edges of town, we find walls around new gated communities and walls around the new homes inside them, or gates at the edges of neighborhoods. The mirror-opposite communities, Lusaka's poor compounds, face barriers of a different sort. As new walls surround elite estates and new or improved roads serve to link them, compound areas are increasingly cut off from pedestrian routes through elite areas and further distanced from the road network. The exception to this is in the recently approved plan for a ring road around greater Lusaka as a part of the latest Master Plan, but if built according to plan this road will displace compound residents in "unauthorized" locations and bisect contiguous compound communities as it slices its way around the city.

Lusaka's formal road grid has always been a spatial manifestation of its coloniality. When one is inside a planned pod, the roads follow a mix of grid and curvilinear logic, with all roads bearing names and street signs. In the unauthorized compounds and in government-planned areas for the poor that have re-informalized, such as Mtendere or Kaunda Square, roads seldom follow a regularized form, or have tarmac, names, or signs. The commercially-produced street map available for purchase at the Manda Hill Mall would be useless for navigating the areas where two-thirds of the city's population resides, since nearly all peri-urban compound areas are actually covered by the map's advertisements. Public transport mini-bus routes dump passengers from every compound into downtown Lusaka—there are no routes which connect from compound to compound across the city east to west or north to south.

These patterns of disconnection and invisibility are not likely to be redressed through the new Master Plan. In its six year gestation period from 2007–13, the Japan International Cooperation Agency's plan for Lusaka built little goodwill in the compounds. In Chibolya and Misisi, for example, residents heard the Lusaka District Commissioner, at a 2012 news conference on the Master Plan, declare that in the Plan's implementation their compounds would be demolished without compensation: "You can't have a city with such types of houses meters away from the Central Business District, those houses will have to go," he remarked (Zambian Watchdog, 2012). While other officials quickly contradicted the DC, and no demolitions had occurred as of June 2013, it remains the case that the planned ring road calls for significant demolition of compound areas and relocation of residents to land to be allocated in Chongwe District, east of the city and far from the existing road grid and the lives and livelihoods of these residents. This is a pattern of planning practice with deep roots in the "structures, attitudes and feelings" of colonial cities across Africa, as cited above, regenerated for the twenty-first century.

Second, despite the evidently successful transition to multiparty liberal democracy in Zambia, it is highly questionable how much of a voice the people from compounds such as the seemingly doomed neighborhoods of Misisi or Chibolya have in any urban development. Indeed, despite the disappearance long ago of racial barriers in the city, many of the real powers or key stakeholders in the city, including major landholders, are foreigners, whites, South Africans, Chinese, or mixed-race Zambians. The barriers which once existed to popular participation in local politics built around race or property were technically wiped away at the end of colonialism (Myers, 2006). Symbolically and almost metaphysically, they still hold sway, perhaps nowhere more starkly than in the office of the (white) then-Vice President of Zambia, Guy Scott; in 2013, Scott was occupying the Lusaka City Council Nursery offices in Woodlands, just beyond the Golf Club, at the old colonial city boundary—the nursery where the seedlings for the colonial Garden City avenue trees were literally grown.

CONCLUSION

Lusaka is just one city, on a continent with more than 50 cities that have populations of at least one million people. There are certainly differences across Africa, in terms of how urban frontiers are changing or not changing. There are differences across the continent in the varying degree of contestation over indigenous claims to land or active state involvement, in the shape and character of informal settlement, in the extent of a private sector or of foreign investment in urban expansion. Although the Chinese have become major players nearly everywhere in Africa, in other countries' cities investment in housing development from Russia, India, Brazil, or South Africa may be more important.

Yet Lusaka does seem representative of some common themes for urban frontiers in Africa in the early twenty-first century. This new, or relatively new, cast of foreign investors is the first theme we see in virtually every major city, even if the particular mix of investors may vary. A second theme has to do with the gap between the arena of expansion propelled forward by such investments and the far less capitalized, yet more prevalent, expansion of informal settlements, often even further away from the peri-urban fringes. Gated communities, closed-circuit television, and high-walled compounds guarded by private security companies show one side of the securitization of Africa's cities; the other side appears in what those gates and walls keep out, in campaigns for "safer cities" bent on securing control over unruly informal areas (Samara, 2011). Such policy campaigns, expanding much like the seemingly ubiquitous and similar plans for controlled neighborhoods such as Silverest Gardens, also indicate a new volume of flows at regional and global scales into which Africa's cities are increasingly tied.

The new scramble for Africa is in some ways, then, leading to a new scramble for African cities. New global flows of investments are transforming urban form across the continent. Some of these transformations appear, on the surface, to be genuinely new. African cities are rife with new housing projects, and these are not just elite gated community projects; middle-income estates and apartment complexes are going up alongside new low-income tenements in some cities. Ultimately, though, the appearance of newness at the gleaming edges of Africa's cities, in shiny malls and glossy elite neighborhoods, seems to be shockingly reminiscent of the late colonial order of many cities on the continent. We see this specifically in exclusionary, income-segregated housing developments, but also in exclusionary planning practices which treat the residents of Africa's "unauthorized" urban areas much as they were treated in the urban development processes of colonial states more than a half-century ago.

I would like to thank Angela G. Subulwa, Mushe Subulwa, Wilma Nchito, and Evaristo Kapungwe for their help and advice with this chapter, even as they bear no responsibility for its shortcomings.

REFERENCES

Akinyele, R. (2009). "Contesting for space in an urban centre: The Omo Onile syndrome in Lagos." In F. Locatelli and P. Nugent (eds) *African Cities: Competing Claims on Urban Spaces*. Leiden: Brill. pp. 109–34.

Ardayfio-Schandorf, E., Yankson, P., and Bertrand, M. (2012). *The Mobile City of Accra: Urban Families, Housing and Residential Practices*. Dakar: Codesria.

Bénit-Gbaffou, C., Didier, S., and Peyroux, E. (2012). "Symposium: Circulation of security models in southern African cities: between neoliberal encroachment and local power dynamics." *International Journal of Urban and Regional Research*, 36(5), pp. 877–89.

Bousquet, A. (2010). "Water and the poor in Nairobi: From water apartheid to urban fragmentation—the case of Kibera." In H. Charton-Bigot and D. Rodrigues-Torres (eds) *Nairobi Today: The Paradox of a Fragmented City*. Dar es Salaam: Mkuki na Nyota Publishers, pp. 121–48.

Bradley, K. (1935). *Lusaka: The New Capital of Northern Rhodesia*. London: Jonathan Cape.

Carmody P. (2011). *The New Scramble for Africa*. Malden, MA: Polity Press.

Carmody, P. and Hampwaye, G. (2010). "Inclusive or exclusive globalization: The impacts of Asian-owned businesses in Zambia." *Africa Today*, 56, pp. 84–102.

Chilufya, L. (2013). "Why China still matters." *Saturday Post* [Lusaka], 15 June 2013, p. III.

Collier, P. and Venables, A. (2014). "Housing and urbanization in Africa: Unleashing a formal market process." In A. Joshi-Ghani and E. Glaeser (eds) *Rethinking Cities*. Washington, DC: The World Bank.

Collins, J. (1977). "Lusaka: Urban planning in a British colony." In G. Cherry (ed.) *Shaping an Urban World*. New York: St. Martins Press.

Cornhill, L. (2013). Personal Communication between author, Mushebe Subulwa, Angela Gray Subulwa, and Loreto Cornhill, sales agent, Silverest Gardens, 14 June 2013, Lusaka.

de Boeck, F. (2012). "Spectral Kinshasa: Building the city through an architecture of words." In T. Edensor and M. Jayne (eds) *Urban Theory beyond the West: A World of Cities*. London: Routledge, pp. 311–28.

Diouck, A. (2013). Author's focus group interview with Alioune Diouck, Mayor of Djidah Thiaroye Kao Commune d'Arrondissement, City of Pikine, Dakar, with Tom Hanlon, Alasse Elhadji, four US undergraduate students and staff members of Partners Senegal, 4 January 2013, Pikine, Senegal.

Fall, A. (2013). Author's focus group interview with Abdoulaye Fall, Vice-Mayor of Thiaroye Gare Commune d'Arrondissement, City of Pikine, Dakar, with Tom Hanlon, Alasse Elhadji, four US undergraduate students and staff members of Partners Senegal, 7 January 2013, Pikine, Senegal.

Grant, R. (2009). *Globalizing City: The Urban and Economic Transformation of Accra, Ghana*. Syracuse: Syracuse University Press.

Hansen, K. (1997). *Keeping House in Lusaka*. New York: Columbia University Press.

HGDC (Henan Guoji Development Company). 2012. *Silverest Gardens: Guoji Dream Town* [brochure]. Lusaka: ZDA- Henan Guoji Development Company, Ltd.

Huchzermeyer, M. (2011). *Tenement Cities: From 19th Century Berlin to 21st Century Nairobi*. Trenton, NJ: Africa World Press.

Kapungwe, E. (2013). Personal communication with Evaristo Kapungwe, lecturer in geography, University of Zambia, 18 June 2013, Lusaka.

Mulenga, A. (2013). "Two die in ZNS operation to remove squatters." *Saturday Post* [Lusaka], 15 June 2013, p. 4.

Murray, M. (2011). *City of Extremes: The Spatial Politics of Johannesburg*. Durham, NC: Duke University Press.

Murray, M. (2013). "Waterfall City (Johannesburg): Privatized urbanism *in extremis*." Unpublished paper presented to the symposium, *Connecting to the City: Belonging, Bordering, and Building Across Time and Space*, 19–20 April 2013. Hartford, CT: Trinity College.

Myers, G. (2003). *Verandahs of Power: Colonialism and Space in Urban Africa*. Syracuse, NY: Syracuse University Press.

Myers, G. (2005). *Disposable Cities: Garbage, Governance, and Sustainable Development in Urban Africa*. Aldershot: Ashgate Publishing.

Myers, G. (2006). The unauthorized city: Late colonial Lusaka and post-colonial geography. *Singapore Journal of Tropical Geography*, 27(3), pp. 289–308.

Myers, G. (2011). *African Cities: Alternative Visions of Urban Theory and Practice*. London: Zed Books.

Myers, G. (2015). "A world-class city region? Envisioning the Nairobi of 2030." *American Behavioral Scientist*, 59(3), 328–46.

Myers, G. and Murray, M. (2006). "Introduction: Situating contemporary cities in Africa." In M. Murray and G. Myers (eds) *Cities in Contemporary Africa*. New York: Palgrave.

Myers, G. and Subulwa, A. (2014). "Narrating Zambia from Lusaka and Mongu." Unpublished paper presented at the annual meeting of the Association of American Geographers, 9 April 2014, Tampa, FL.

Nchito, W. (2013). Personal communication with Wilma Nchito, lecturer in geography, University of Zambia, 26 June 2013, Lusaka.

Ngau, P. (2012). Personal communication with Peter Ngau, professor of urban planning, University of Nairobi, 21 July 2012, Nairobi.

Potts, D. (2004). "Regional urbanization and urban livelihoods in the context of globalization." In D. Potts and T. Bowyer-Bower (eds) *Eastern and Southern Africa: Development Challenges in a Volatile Region*. Harlow: Pearson Education.

Pullan, R. (1986). "The biogeography of Lusaka, with special reference to trees and shrubs." In G. Williams (ed.) *Lusaka and its Environs: A Geographical Study of a Planned Capital City in Tropical Africa*. Lusaka: Zambia Geographical Association, pp. 278–89.

Rakodi, C. (1986). "Colonial urban planning in Northern Rhodesia and its legacy." *Third World Planning Review*, 8, pp. 193–218.

Samara, T. (2011). *Cape Town after Apartheid: Crime and Governance in the Divided City*. Minneapolis: University of Minnesota Press.

Tait, J. (1997). *From Self-help Housing to Sustainable Settlement: Capitalist Development and Urban Planning in Lusaka, Zambia*. Brookfield, VT: Avebury.

Trefon T. (2009). "Hinges and fringes: Conceptualizing the peri-urban in Central Africa." In F. Locatelli and P. Nugent (eds) *African Cities: Competing Claims on Urban Spaces*. Leiden: Brill, pp. 15–36.

UN Habitat (2010). *The State of African Cities 2010: Governance, Inequality and Urban Land Markets*. Nairobi: UN Habitat.

Zambian Watchdog (2012). "PF plans to demolish Lusaka compounds now a reality." *Zambian Watchdog*, 24 May 2012. Available at: http://www.zambianwatchdog.com (accessed 18 October 2013).

4

Imagining and Staging an Urban Border: The Role of the Netherbow Gate in Early Modern Edinburgh

Giovanna Guidicini

The significance of the presence of a wall in relation to an urban environment has been extensively discussed, emphasizing its role as a dividing element as well as its representation as a join between different realities. At a city gate passage is negotiated, denied, granted, and at times conquered through violence. This essay studies one city gate in particular, the Netherbow in Edinburgh, during the early modern period, at which time it was a public space heavily loaded with political, social, and cultural undertones. From the earliest records of its existence to its demolition in the nineteenth century, and beyond that to current times, the Netherbow has represented a location where the city's identity was negotiated among those approaching it from within and from outside, particularly once new expectations were created by the Union and the modern era.

THE NETHERBOW AND THE SIGNIFICANCE OF CIVIC SPACES AND URBAN BORDERS

The famous map of Edinburgh from 1647 by James Gordon of Rothiemay is one of the earliest reliable illustrations representing the organization of the burgh town. The Netherbow is the city gate visible at the eastern end of the royal burgh of Edinburgh, a densely built–up area to the left of the map, dominated by the Stewarts' castle (Harris, 1996, p. 455). To the right of the map lies the Canongate, a neighboring but distinct burgh separated from Edinburgh by a tall wall and by the Netherbow gate itself.

The Netherbow, which in 1369 was called *arcus inferior*, was matched by the Upper Bow or Over Bow standing at the upper end of the burgh, which was later superseded in its role of frontier by the West Port (Turner, Stevenson and Holmes, 1981, p. 12). The Netherbow and the West Port represented the two main access points to the city; the Netherbow, in particular, guarded one of the few carriageable entrances into town, as well as the road to the nearby port of

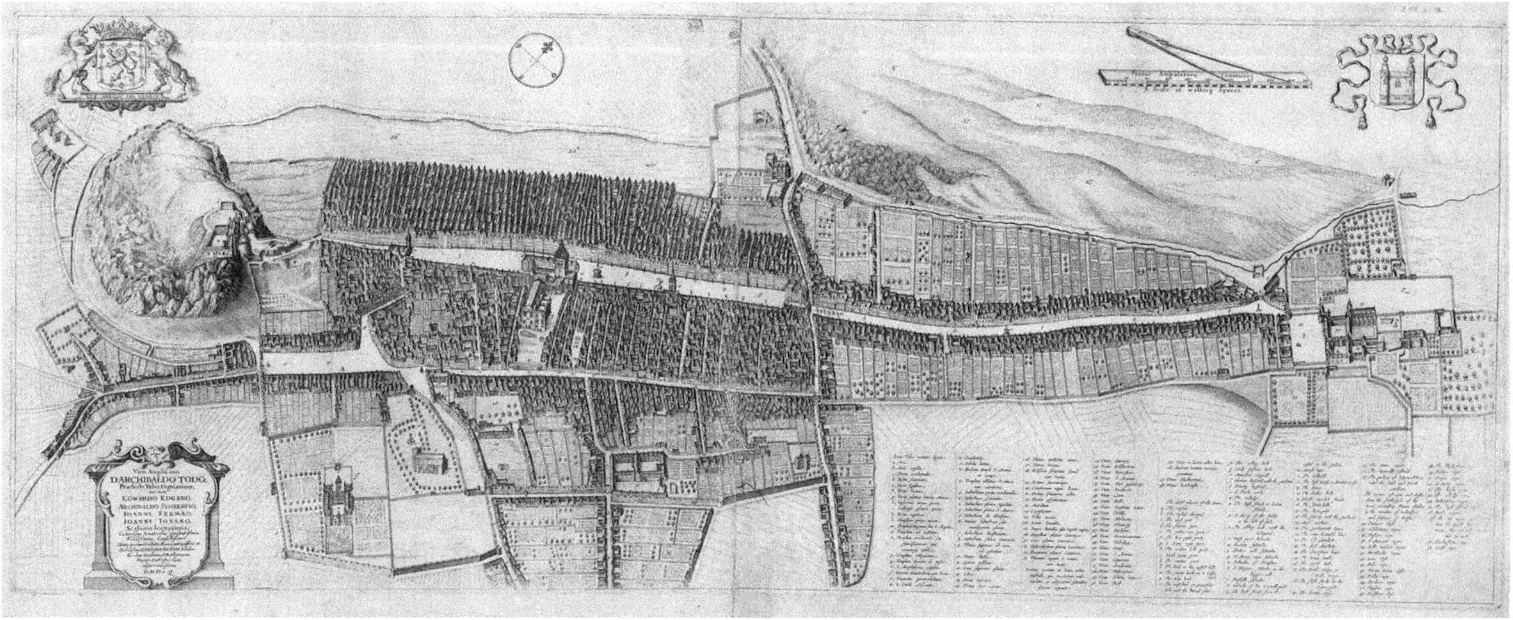

4.1 James Gordon of Rothiemay, *Edinodunensis Tabulam*, Amsterdam ca. 1647. Reproduced by permission of the National Library of Scotland.

Leith, which represented a major military defense. The so-called English spy's map drawn by English agent Richard Lee in 1544 shows the city from the point of view of an advancing army, and demonstrates particular care in the representation and positioning of the Netherbow. This gate had an emphatically martial appearance: it was rebuilt in 1503, damaged in 1544, and rebuilt in 1573, probably to resemble the Portcullis Gate at the entrance to Edinburgh Castle (Dennison and Lynch, 2005, p. 30). It was subsequently rebuilt again in 1611 in an elaborate castellated style, with crenellations and gun loops. However, the Netherbow's role goes beyond that of military defense. I argue that it represented one of the symbols of the urban community's identity, and the contested ground on which this identity was constantly being adjusted and renegotiated.

According to Max Weber (1958, pp. 80–81):

> *a full urban community ... must display a relative predominance of trade–commercial relations ... [and] 1. a fortification; 2. a market; 3. a court of its own and at least partially autonomous law; 4. a related form of association; and 5 ... an administration by authorities in the election of whom burghers participated.*

As suggested by Peter Arnade (2013), city gates, market squares, and other public spaces where community activities took place, are the locations where civic identity was expressed, authority was legitimized, and privileges were negotiated during the Early Modern period. Spyro Kostof (1992, pp. 124–5) notices here the potential for an urban paradox, as ruling agencies try to order and control the intrinsically unregulated social spaces where citizens recognize themselves as a community. If urban spaces become the physical representation of the city's identity and communal values, in line with Henri Lefebvre's ideas on the causative power of space, then altering and unsettling the former has profound consequences on the latter. Lefebvre (1991, p. 47) argues that the architectural code which allows us to decipher reality becomes the parameter by which reality is constructed. This "would allow the organization of the city, which had been several times overturned, to become knowledge and power." The mutable configuration of cities becomes then a true representation of the Early Modern economic, political, and social transformations. The Palazzo Vecchio was appropriated and altered by the Medici family in Renaissance Florence, becoming a visual justification of the change from Republic to Oligarchy (Strong, 1984, p. 87; Testaverde, 1980, pp. 77–9). In *Power/Knowledge*, Michel Foucault (1980, p. 148) discusses the transition from architecture as "the need to make power, divinity and might manifest" to architecture becoming "the disposition of space for economico-political ends." Foucault (1980, p. 149) explains how "from the great strategies of geo-politics to the little tactics of the habitat," the control of space and the exertion of power inevitably overlap. Similarly, Pierre Bourdieu (1977, pp. 89–95 and pp. 159–67) in his studies of a Berber group shows how the dominant groups in a community maintain their power by shaping spaces that reinforce their authority and embody social order. Inferior groups are compelled to remain in their subservient position until they obtain "the material and symbolic means of rejecting the definition of the real that is imposed on them" (Bourdieu, 1977, p. 169) by the dominant orthodoxy.

Gaining control of a city's walls and gates—the boundaries that regulate and differentiate the inside from the outside—questioned and destabilized the community's perception of citizenship, safety, rights, and privileges. The destruction or alteration of a rebellious city's urban spaces by the victorious sovereign represented a permanent memento of the city's inability to defend its buildings, as well as the privileges those buildings embodied (Arnade, 2013, pp. 739–40). After all, as Isidore of Seville emphasized, "*urbs ipsa moenia sunt*" (Isidore of Seville, 1911, XV, ii.1)—the city is made by its defensive walls—with the "*civitas*," the community, living within them. Kostof (1991; 1992) argues that the dialectic between the inside and the outside identities of an urban fortification depicts the story and characteristics of the community living within it. Michel de Certeau (1988, p. 117) famously states that "space is a practised place," emphasizing the role of inhabitation and its use in the creation of spatial stories. Urban boundaries in particular narrate stories of "isolation and interplay" (de Certeau, 1988, p. 123), both literary and figurative. A boundary distinguishes a subject from its surroundings, mapping patterns of behaviors, obligations, and interactions; it defines a network of movements, differentiating the home within from the journey beyond (de Certeau, 1988, pp. 122–5). The different spaces created by the threshold interact with each other, creating "narratives of exclusion or inclusion, control or comfort ... which become inscribed in the daily practice of their use and in turn also influence that use" (Stalder, 2009, p. 75). Discussing the stories created by this constant interplay of forces at and around the Netherbow will become the key to understanding the significance and role of this urban frontier.

The urban gate is a politicized space, a stage where the city has the right to negotiate with external forces, and with the state authority itself (Arnade, 2013, pp. 739–41). The significance of walls and city gates as representative of civic identity is particularly evident in the case of Edinburgh, which was established as a royal burgh by King David I (1084–1153); this meant that the crown was the city's only feudal superior. The city's rights and privileges, as well as its obligations, were stated by the king in front of witnesses, and recorded on a charter or act (Keith, 1913; Mackay Mackenzie, 1949, pp. 62–84). Among other financial privileges, such as a right to hold a market, Edinburgh was granted the right to build a defensive perimeter of stone walls around the city (Cullen, 1988, pp. 1–4; Turner Simpson, Stevenson and Holmes, 1981, pp. 18–19). Built with the king's consent, the fortification of the burgh did not create a competitive political entity but was an expression of monarchical authority. The construction of walls implied and represented the presence of a specific political and social structure; it represented a pact between the prince, who gave the town the right to build and to raise revenues for that purpose, and the community that was willing to finance, organize, construct, and defend the walls (Tracy, 2000). Within this walled perimeter, a florishing mercantile community led by powerful guilds "remained dedicated to the mission of preserving the burgh as a fortress of economic privilege" (Lynch, 1981, p. 14). The cohesive role of commercial identity was physically expressed by the encircled urban space: the extremely tall, adjoining tenement buildings demonstrate the significance attributed to living within the urban perimeter rather than in the more spacious suburbs. As late as the early modern period, residency was one of the requisites—although not always strictly enforced—of burgess-ship (Allen, 2011, p. 427–9). The burgh's main public buildings on the High Street also expressed the city's right to self-governance: the Market Cross embodied the city's right to hold a market, the Tolbooth represented civic administration and justice, St Giles Kirk its religious authority, and the Trons were market locations as well as authorized weighing posts. The role of the Netherbow is, then, not only occasionally to defend the city from military aggression, but to confirm daily and reiterate the burgh's rights as inseparable and deriving from the extent of its civic space, in opposition to a distinct outer space governed by different rules. The role of an urban portal as a guardian between worlds is perhaps easier to acknowledge when the density of the urban centre is juxtaposed with the openness of the natural world (Liddiart, 2005). In the late fourteenth-century tale *Sir Gawain and the Green Knight*, Sir Gawain, worn out by his adventures in the realm's wilderness, passes through a sizable border to get access to the longed-for safety of a castle; "the bridge was securely lifted, the gates locked fast; the walls were well arrayed; no wind blast did it fear" (Neilson, 1999, p. 17). Britt C.L. Rothauser (2009, pp. 248–9) suggests that the medieval walled city can be compared to Manfred Kusch's concept of walled garden as a "rationally controlled system surrounded by an often amorphous wilderness" (Kusch, 1978, p. 1). The burgh of Canongate, however, provided a different representation of otherness.

King David established the burgh of Canongate as a service community for the Abbey of Holyrood; he had founded it in 1128 to provide goods and services to the Abbey, and later to the Stewart's own Palace of Holyrood (Dennison, 2005, pp. 5–12 and pp. 28–31). As a subservient burgh to the Abbey of Holyrood, Canongate was not entitled to Edinburgh's defensive and commercial privileges. Its boundary walls were not defensive, but only apt to "answer ordinary municipal purposes" (Wilson, 1848, II, p. 57; Grant, 1880–83, III, p. 2). It had no right to hold a market, and did not enjoy Edinburgh's right to international trade; Canongate's craftsmen were allowed to work only within the perimeter of their own burgh (Allen, 2011). Their relationship with the sovereigns was also different: while Edinburgh became more and more the centre for royal government, Canongate and Holyrood Palace were the seat of the royal court (Dennison and Lynch, 2005, pp. 35–6). The Netherbow stood guard to those differences, and physically expressed the boundary between the different urban spaces and identities.

As an opening in an otherwise impenetrable wall, a gateway is permeable and accessible by nature, representing a link between the realities it delimits. Sennet (2004) defines a wall as a structure to inhibit passage, not unlike a cell membrane, but also underlines its porous character and its ability to be crossed under certain conditions. Du Certeau (1988, pp. 126–7) argues that dividing elements—the river, the door, the picket fence—represent as much a frontier as they do a crossing, as they can be passed, opened, or glanced through. Gates need then to compromise military impassability and functionality as an urban access, with the additional requirement of architectural monumentality to represent the community appropriately to those approaching from the reign's highways and seeking admittance (Kostof, 1992, pp. 36–7; Reyerson, 2000). In the following section I discuss the different roles of the Netherbow through an analysis of the historical events and narratives that involve it, to understand how the gate's different identities came together to create a coherent early modern urban frontier.

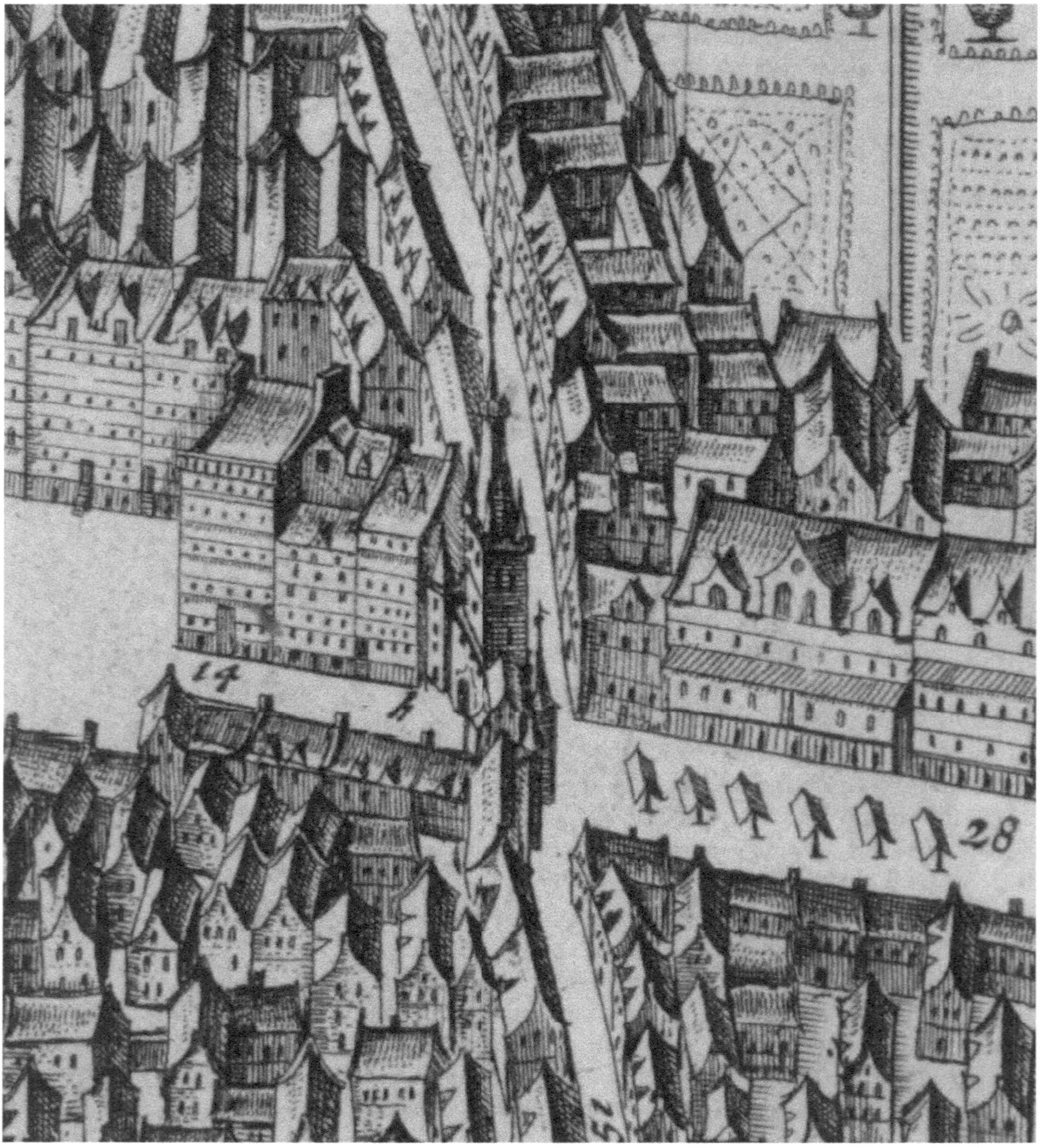

4.2 James Gordon of Rothiemay, *Edinodunensis Tabulam*, detail, Amsterdam ca. 1647. Reproduced by permission of the National Library of Scotland.

4.3 *The Netherbow Port*, from the Canongate. From an Etching by James Skene of Rubislaw, published in James Grant's *Old and New Edinburgh* (1890) Vol. II, p. 201. Reproduced with acknowledgement to Peter Stubbs: www.edinphoto.org.uk.

FRIENDLY MEETINGS: NETHERBOW GATE AS OCCASION FOR DIALOGUE, NEGOTIATION, AND EXCHANGE

As a passable border between two different communities, the Netherbow was frequently crossed by both sets of burgesses for a variety of economic, politic, and social reasons. Although not entitled to a market within their own burgh until possibly the late sixteenth century, the Canongate burgesses did have the right to use the Edinburgh market freely, and benefited from being in Edinburgh's market catchment area, as they would legitimately cross the border on their way there (Dennison and Lynch, 2005). The Netherbow was also crossed daily, often illegally, by Canongate traders who disregarded urban borders and spatial significance for personal gain; as reported indignantly by Edinburgh's incorporation of the tailors in 1584, their Canongate counterparts "daylie cumis within the fredome of the samyn and takis furth wark, shaipin and unshaipin, pertening to the burgessis and friemen of this burgh" (Marwick, IV, 1882, p. 367; Allen, 2011, p. 429). Edinburgh burgesses also crossed the urban border for reasons of convenience; until the 1540s, the Edinburgh guild of the hammermen frequently used the great hall of the Dominican friary, located outside and to the east of the Netherbow, near the Canongate's southwest corner, as a meeting hall, for witnessing charters, and for the annual exchequer audit (Dennison and Lynch, 2005, p. 25; Foggie, 2003, pp. 83–5). The physical border between the two burghs was actually rather blurred, and inevitably some geographical overlap occurred. In 1571, during the Marian civil wars, the king's party held the so-called "creeping parliament" at the Canongate in the house of a William Oikis, which the chronicler notices was "within the freidom of Edinburgh, albeit the samyne wes nocht within the portis thairof" (Scott, 1833, I, p. 214; Dennison, 2005, pp. 48–9). Generally, the Edinburgh burgesses established themselves outside the walls for economic reasons, such as lower taxation, absence of duties for town watch and militia, and better living conditions (Allen, 2011, pp. 429–30). Although these outland burgesses abandoned Edinburgh's civic space physically, they still claimed the freedoms that were enjoyed within its perimeter. Catholic burgesses of Edinburgh also temporarily removed themselves from their kirk's jurisdiction after the reformation, and relocated to the other side of the Netherbow (Lynch, 1981, p. 43), as did Presbyterian burgesses of Edinburgh who changed place of worship rather than adopting the practices enforced by King James VI's five articles of Perth (Stewart, 2006, p. 193; Haddington, 1837, p. 633). A different reason to cross the urban border would be to take refuge in Canongate's sanctuary' area, at the girth cross, where the right of sanctuary was granted to debtors, making it an alternative to bankruptcy for ruined burgesses (Dennison, 2005, pp. 40–43).

This panorama shows the Netherbow as an extremely flexible and at times muddled urban border, being crossed frequently, both legitimately and illegitimately, by a variety of people. Personal convenience seemed a strong factor in prompting burgesses to ignore the limitations and expectations placed on their conduct by their belonging to one or another urban precinct. While the economic advantages represented by the Netherbow were both defended and coveted, the responsibilities and obligations that came with it and the burgh's right to interfere with its citizens' lives were resented. The Netherbow represented, then, more a permeable access-point than a clear-cut separation, and it was seen as both a potential way into a world of privilege, and also a way out for exasperated citizens looking for alternatives.

A blurred and negotiable urban border often promotes the creation of a grey area, a buffer zone where rules are uncertain and danger is ever–present (Alpar Atun and Doratli, 2009). The dissolution of the extramural Dominican estate after the reformation promoted the creation of an unregulated suburb, the Pleasance. This row of houses built on the friars' farmland opposite the urban wall became known as "Thief Row" (Allen, 2011, p. 425), suggesting that the extramural location and in-between status implied a certain degree of lawlessness. The presence of many dangers lurking in the shadows is confirmed by records made at the time. For example, in 1708 records show that a James Baird waited to settle his drunken fight with Robert Oswald until their coach left them outside the Netherbow port late at night, whereupon Baird unfolded his sword and mortally wounded his companion before disappearing into the darkness (Grant, 1880–83, VI, p. 319). treacherous enemies could lie in wait in the row of houses outside the Netherbow, ready to invade and threaten the civic order: during the Lord Regent's siege of Edinburgh in 1571, soldiers came from Leith "during the night and conceal themselves in the closes and adjoining houses immediately without the Nether Bow port … ready on a concerted signal" (Wilson, 1848, II, p. 57). In 1745, a party of Cameron assailants "were ordered to place themselves on each side of the gate," where a substantial support was stationed "in deep silence" at nearby St Mary's Wynd (Scott, 1841, II, p. 426).

Gaining admittance and passing through the urban gateway marks the visitor's changed status from possible aggressor, or at best unknown entity, to harmless visitor or even friend. The altering characteristics of the urban border are highlighted by the symbolic use of the Netherbow in a tale recorded in *Memorials of Edinburgh* (Wilson, 1848, II, pp. 61–2). the story tells of a delegation of Edinburgh's magistrates who entered into parlay with a group of threatening Moor pirates at the Netherbow, where an agreement was reached with the pirates' chief requiring the Provost to hand over his son to spare the city. However, on hearing that the Provost has a sick daughter, the pirates' chief softens, offers to heal her and is in turn given access to the city to visit her. He refuses the invitation, and asks instead that the girl join him and his party in the Canongate. The girl is subsequently returned to her father in good health, and according to a second narration ended up marrying her benefactor and settling with him in the same house in the Canongate where she had been healed. The Moors' chief reveals himself as one Andrew Gray, initially captive, then confederate and finally leader of the pirates, who had vowed to revenge himself for early wrongs instigated by the magistrates of Edinburgh. A change of heart prompted by his marriage to the Provost's daughter made him abandon his planned revenge, but he settled in the Canongate because "he had vowed never to enter the city but sword in hand; and having abandoned all thoughts of revenge, he kept the vow till his death, having never again passed the threshold of the Nether Bow port" (Wilson, 1848, II, 62).

The Netherbow is presented as a place where outsiders seek admittance, and the insiders pose conditions, whose terms are discussed. In many literary works, such as the knight's tale's narration of Theseus' approach to Athens, the liminal space outside of town acts as a buffer zone, protecting urban harmony from conflict (Jost, 2009, p. 377). Spatial conventions are strictly respected while negotiations take place, with each party sticking to their allocated spaces, discussing with counterparts across the physical barrier. While access is obviously refused to a party of invaders, once the Moor declares he is willing to cure the Provost's daughter he is invited to cross the threshold as a guest. Each time the girl crosses the threshold, her status changes: from coveted upper class girl to potential hostage, from sick to healed, and finally from maid to married woman. The pirate chief also undergoes momentous changes at the Netherbow; from threatening Moor to friendly Moor, to his original identity as Andrew Gray, to established married citizen—but interestingly he does not cross the city border, afraid of the Netherbow's power to force him to enact his own vows and turn him into the avenging assailant he swore he would be. At the Netherbow, a visitor's altered status is acknowledged, and change becomes visible and public: in 1662 a group of newly consecrated bishops walked in procession—formally dressed and in full view of the whole city that witnessed their changed status—from the Netherbow to the Parliament, where they were honorably received (Wilson, 1848, II, p. 55). Notwithstanding its urban surroundings, in this account novel the Netherbow does partially fit the *topos* of the gateway between the security of urban life and the mystery and dangers of outside wilderness. Also, in popular European folklore wild men and moors represented the foreign, the savage and the unknown, a potential threat to community rules (Bernheimer, 1952; Mcmanus, 2002, pp. 87–9, pp. 194–7). The urban frontier is then the space where extraordinary events—meaning extra-ordinary, out of the normal conventions and rules of daily life—take place. Here, passing the Netherbow, mysterious drops of fresh blood from the desiccated head of Gurhrie permanently stained the coach of his nemesis General Middleton (M'Gavin, 1828, p. 248). The unexpected and the mysterious—an enigmatic enemy Moor with an "elixir of wondrous potency" (Wilson, 1848, II, p. 61)—meet the conventional, the sickened young citizen of a crowded city often stricken by contagious illnesses, and miraculously reveals himself as benevolent savior.

The prodigious transformative power of urban thresholds is also evident in the role of the city gates and other civic landmarks during triumphal entries. Early modern urban processions in Edinburgh touched all of the city's administrative and commercial public structures; the rulers were welcomed at the West Port, proceeded through the Over Bow and arrived at the Butter Tron. They then moved down the High Street into the core area of Market Cross/St Giles Kirk/Tolbooth, down again past the Salt Tron and exited the city at the Netherbow. From there the procession continued through the Canongate and reached the Palace of Holyrood. These locations represented specific aspects of Edinburgh's identity, and the privileges and rights that the ruler's ancestors had granted her. While a sovereign conquering a city was a humiliating appropriation of politically significant urban spaces, during triumphal entries here these meaningful locations are emphatically lent to the rulers as an act of loyalty in celebration of their majesty (Arnade, 2013). They become relational spaces where two different sets of values and social orders apply—courtly and civic—and it is where the different people they represent come into contact (Hirschbiegel and Zeilinger, 2009). In particular, the gates represented a city's right to allow or deny access to visitors, and to defend itself against enemies; so willingly opening the urban perimeter to the visiting monarch was a sign of trust and loyalty (Pepper, 2000).

In Edinburgh, the ruler was emphatically received at the West Port as the burgh's munificent patron, where he would receive the city's keys and other symbols of loyalty and welcome (Guidicini, 2011). At the end of the ceremony, the Netherbow bid goodbye to the departing monarchs, showing the city's wishes and expectations for the continuation of their rule. The triumphs for James VI in 1579 and for Charles I in 1633 were staged with astrological themes, which represented the favorable conjunction of planets at the time of the King's birth (Colville and Thomson, 1825, p. 179; Anon., 1633). At the Netherbow in 1590, Queen Anne of Denmark saw the spectacle of a serene ruling couple surrounded by ministers and courtiers, wishing her own marriage stability and happiness (Craig, 1828, p. 42). The dragon shown here to Mary Queen of Scots in 1561 also represented a prediction for the queen's future, as it was interpreted either as the beast challenging the biblical maiden giving birth to a great king or as the beast from the abyss accompanying the whore of Babylon, a dark prediction of the consequences of the queen's Catholicism (MacDonald, 1991, p. 107). The symbolically charged messages delivered throughout the entrances aimed to transform the ruler into the benevolent patron the burgh wished for. The Netherbow represented the climax of this transformation, and released into the outside world a persuaded and captivated monarch, who would protect the interests of his capital.

The lack of official triumphal stations outside the Netherbow underlines the different characteristics of the two urban spaces, but the occasional spillover of the ceremony into Canongate challenges the perception of the Netherbow as both urban border and farewell location.[1] For example, in 1560 a delegation of burgesses from Edinburgh went past the Netherbow accompanying Mary Queen of Scots to her rooms in Holyrood Palace, where they presented her with a gift (Scott, 1833, I, p. 69; Guidicini, 2011). In 1579, James VI heard a speech at the Canongate Cross regarding making the mass illegal; the lack of official Edinburgh records on this event suggests it to be an unauthorized extra mural celebration (Mill, 1924, p. 194).[2] In 1617, in the palace's inner courtyard, James VI accepted

---

1 It is worth noting that most—although not all—of the accounts of urban ceremonies come from official sources and might present Edinburgh's point of view more than Canongate's. Unofficial and unauthorized celebrations might have taken place outside of the Netherbow but gone unrecorded by Edinburgh's chroniclers.

2 From Mill's transcription of Johnston's MS. History of Scotland. Adv. Lib. Hist. MSS. 35.4.2.

the homage of a group representing the recently founded Edinburgh College (Historical Manuscript Commission, 1894, p. 20). However, the symbolic importance of the Netherbow as a frontier was emphatically reintroduced in 1617, when at the Netherbow King James VI knighted the Edinburgh Provost Wlliam Nisbott before the Edinburgh dignitaries passed over to "the Baillies of the Cannongait … with their company" (Historical Manuscript Commission, 1894, p. 20), who escorted him to the Palace of Holyrood. This event conveys a sense of acknowledged equality between the two communities, and the complexity of the event suggests some form of coordination between the burghs' authorities: the Netherbow acted both as an exit from Edinburgh and as an acknowledged entry into Canongate. The Canongate's connections with the royal court and the acquisition of the smaller burgh by Edinburgh in 1636 might have promoted feelings of collaboration over competition (Dennison and Lynch, 2005, pp. 39–40; Dennison, 2005, pp. 49–50).

While triumphal entries placed an emphasis on civic spaces and urban borders, parliamentary and religious processions did not emphasize the Netherbow's role as frontier. Parliamentary processions, or the Riding of the Parliament, marked the opening and closing of the Scottish Parliament; the riding group included representatives of the monarch and of the estates, and escorted the honors of Scotland from the Palace of Holyrood up the Canongate, past the Netherbow and up the High Street to the Parliament house (Mann, 2010, pp. 132–3). The burgh's perimeter was not relevant during this supra-local ceremony, in which Edinburgh acted as the national capital where monarchy and the country at large came together. Nevertheless, the burgesses themselves used the Netherbow to determine the spatial extent of their involvement. During the Ride of the Parliament in 1633, "at the Nether Bow, where he entered the bounds of the city, the king was saluted by the provost, who attended him closely the rest of the way" (Chambers, 1858, II, p. 66). In 1594, the citizens refused to make way for the king's own guard of horsemen, "by reason of their priviledge to guarde the king's person in tyme of parliament, till he depart the toun" (Calderwood, 1844, V, pp. 329–30), since the burgh's area was defined in an article of law that forbade the shooting of firearms as "within the ports" (Calderwood, 1844, V, p. 330). The overarching symbolism of religious processions through town also superseded the Netherbow's role as a civic border; in John Knox's records of the yearly St Giles procession from St Giles to the Canon Cross and back, no particular ceremony or event is recorded in connection with this double passing of the Netherbow (Knox, 1831, p. 88).

The Netherbow represents the frontier of Edinburgh's freedoms, and the border between a privileged inside and an undetermined outside. The Netherbow has proved itself to be a permeable and flexible border, where significant exchanges—of ideas, goods, people and civic significance through spatial overspills between inside and outside takes place—both legitimately and illegitimately. For the troubled burgess, the mysterious pirate and the visiting monarch, the Netherbow is a source of change and of significant personal alteration. However as a permeable barrier the Netherbow can also let danger in: its role as a vehicle of violence and vulnerability will be discussed in the following section.

HOSTILE MEETINGS: NETHERBOW AS A SITE OF AGGRESSION AND VIOLENCE

To people with illegitimate business, the Netherbow represented a physical obstacle, as the legitimacy of the passers–by's business and movements would here be checked by booths and sentinels. This scrutiny would be avoided through deceit and trickery: in 1689, convicted Lord Burleigh was discovered at the Netherbow while trying to escape his sentence by smuggling himself out of the walls hidden in a box (Grant, 1880–83, I, p. 127). Again in 1742, two chairmen were discovered here trying to bring in a dead body—a highly sought-after item, probably on the way to a dissecting table—in a sedan chair (Adams, 1972, p. 19). A more complex relationship, with the Netherbow as guardian of rightful civic behavior, took place in 1567 in connection with the now infamous plot to murder Queen Mary's husband Lord Darnley at Kirk O' Field. The use of civic space and urban borders by the plotting party in these circumstances is quite telling. On the night of the murder, Bothwell returned to Holyrood with the Queen's train from nearby Kirk O'Field, before sneaking out of the palace's gardens and through the South Gate in the small hours, disguised as a servant and with a group of associates. After being let in at the Netherbow under a false identity, they headed for Darnley's house (Mignet, 1851, I, p. 267). Their way back after the explosion is more tortuous, and is explained in detail as "they ran down to the Cowgate, through the Blackfriars gate, and ascending by different closes, crossed the High-street to a broken part of the town wall in Leith Wynd which Bothwell was unable, or afraid to leap" (Laing, 1819, II, p. 9). Bothwell's legitimate activities in town take place in plain sight and on the main roads, crossing gates and borders with the authority of a respectable courtier in the Queen's train. On the other hand, his nocturnal escapade, especially in the commotion after the explosion, sees him thinking and moving like a criminal, rushing in disguise through tortuous routes of back alleys and narrow lanes, avoiding public spaces and main roads. The most challenging space of all is now the unavoidable and alerted Netherbow, which stands between Bothwell's party and the safety of the Palace. By planning to get out of the burgh by a broken section of the wall, Bothwell is trying to avoid the gate's inquisitive and challenging gaze, finding an alternative and unofficial way out of the guarded civic space. Interestingly, Edinburgh's frontiers stand fast and even the broken wall proves too hard to climb: the Netherbow is confirmed as effective city guardian when the party is forced to go through it with the help of a concerned and observant gate keeper (Finlayson Henderson, 1905, II, pp. 436–7).

The Netherbow stood strong against criminal and unacceptable behaviors seen as incompatible with civic life, and here physical exclusion of the culprits from the urban precinct was enforced. Ordinances were frequently set in place for the removal of beggars and other undesirable unsightly people from within the urban precinct for urban ceremonies and triumphal entries (Anon., 1822, p. 61).[3] Gateways become the markers separating the ideal city created for the ruler—prosperous,

---

[3] Noted as "*Notices from the Records of Privy Council, relative to the Reception of the King, 1617*," December 24, 1616.

clean, well-organized, and beautifully decorated—from the unappealing reality of day-to-day life in a metropolis. By refusing to abide by the burgh's law, criminals, delinquents and traitors had placed themselves metaphorically and physically outside of the community of honest citizens. "Upon the Netherbow was a spike of iron, upon which the heads of traitors and others were exhibited" (Chambers, 1836, p. 329); the offenders' remains were removed from the civic spaces, and placed to guard the urban perimeter as a warning to those who approach.

While urban executions took place within the walled perimeter at the Grassmarket, a large square ensuring maximum visibility, military criminals belonging to the king's militia were executed at the Leith Links, outside the city walls (Grant, 1880–83, IV, p. 231; Grant, 1880–83, VI, p. 263). This underlines the army's justice being seen as not subject to the same rules and spatial bounds which regulated the burghs and its inhabitants (Mackillop, 2011, pp. 175–6). The army's otherness to the burgh's civic spaces is evident in how they dealt with compulsory conscription in town: the burgh's magistrates themselves rounded up the eligible men—often petty criminals and troublemakers—within the urban perimeter and handed them over to army officials at a meeting tactfully arranged outside the Netherbow (*The Scots Magazine*, 1755, pp. 154–5).

While removing undesirable people did protect the integrity of the community, the Netherbow was not always successful in defending the urban frontier. In May 1544, the civic authorities failed to bargain an honorable and safe surrender with an invading English army. The magistrates' offer of the keys of the city acknowledged the English army's superiority, but it also assumed some form of negotiating power on the part of the burgh and the right to demand safety for the urban spaces the English are accessing. On the contrary, the English commander's haughty refusal marked his aggressive intentions towards the city and its inhabitants; the Netherbow was "blew open by dint of artillery" (Grant, 1880–83, I, p. 43), an intentional violation of the urban frontier that anticipated extensive pillaging and destruction (Arnot, 1816, pp. 14–15).

While the destruction of a gate was seen as an acceptable act of war that could prevent the city from posing a threat again, it also implied a lost source of revenue and damage to one's newly acquired property (Arnade, 2013). The Netherbow was more often conquered through cunning than brute force; for example, in 1571 a party of James VI's soldiers disguised as millers leading loaded horses planned to hold the gates open with their meal sacks long enough for their comrades to rush through the gate. Observant passer-by Thomas Barrie revealed the presence of a party of concealed soldiers to the gate's watchmen, and the enemy was repelled (Scott, 1833, I, p. 240). Furthermore, during a Jacobite attack on the city in 1745, one of Lochiel's men tried to gain admittance at the Netherbow disguised as the servant of an English officer "in a riding coat and hunting cap" but the sentinels refused to open the gate for him (Scott, 1841, II, p. 426). The Highlanders had several barrels of gunpowder with them to blow up the gate, but decided instead to lie in wait and stormed through the gate when a carriage passed out (Kay, 1842, I/II, p. 419). In 1745, losing the Netherbow to the Jacobites was instrumental to losing control of the city, and the accounts from the city authorities and of the

Castle garrison on how it happened blame each other for carelessness (McLynn, 2003). The importance of the moment was also well understood by the entering Highlanders, who make a parade of the event: "the whole clan Cameron advanced up the street, with swords drawn and colors flying, their pipes playing 'We'll awa to Shirramuir, And haud the Whigs in order'" (Grant, 1880–83, II, p. 325). The bagpipers playing "in spirit-stirring tones" (Kay, 1842, I/II, p. 385) in the civic spaces celebrated the taking of the port and of the whole city, "while the magistrates retired to their houses, aware that their authority was ended" (Grant, 1880–83, II, p. 325). Similarly in 1582, exiled Minister John Durie was welcomed back into the burgh by a cheering crowd, and "at the Netherbow they took up the 124 Psalme, 'Now Israel may say,' &c" (Calderwood, 1849, VIII, p. 226). Psalms and bagpipes, anthems of victory and battle tunes move the hearts of those singing and hearing it, emphasizing the crossing of the urban frontier.

The Netherbow is not a neutral background, but a politicized object which takes the part of those who are controlling it, declaring and embodying the city's affiliations and sympathies. When in 1654 General Monk came to Edinburgh to declare Oliver's union of England and Scotland "the provost and bailies in their scarlet gowns met him at the Nether Bow Port, the hail council in order going before them" (Chambers, 1858, II, p. 225). In contrast, in 1715 the citizens of Edinburgh shut the Nether Bow to William Mackintosh of Borlum and his army, taking a clear stance against the Jacobite cause (Chandler, 1679, II, pp. 300–301). Again in 1736, the part the Netherbow gate played in aiding an angry mob gaining control of the urban space—and the perceived misuse of the city's right to fortify and defend itself—almost led to the gateway's destruction. Since this episode, known as the Porteous incident, is both complex and politically relevant, it deserves closer scrutiny here.

The story is as follows (Parliament, Great Britain, 1812, pp. 187–319):[4] in September 1736, while a Captain Porteous of the City Guard was imprisoned in the Edinburgh Tolbooth awaiting trial, an angry mob took control of the town by closing the city gates. In particular, the barricading of the Netherbow efficaciously prevented the king's army, quartered in Canongate, from entering the city and intervening, and the mob successfully assaulted the jail and summarily hanged Porteous. Alarmed that this violent demonstration of defiance could represent the beginning of uprisings by Jacobite supporters and threaten the good management of Scotland, in February 1737 a Parliamentary enquiry condemned the episode and discussed a punishment. Given the impossibility of identifying the individuals composing the mob, a Bill was proposed to imprison and disgrace the negligent Provost, abolish the town guard, and for taking "away the gates of the Nether Bow Port of the said city, and keeping open the same" (Parliament, Great Britain, 1812, p. 189). Destroying the Netherbow was clearly a punishment intended for the whole urban community and it addressed the symbolic importance of the Netherbow as an urban landmark.

---

[4] Listed as *Proceedings in the Lords Relating to the Murder of Captain Porteous: And the Bill Against the Provost and City of Edinburgh*.

The heated discussion during Parliament between the MPs supporting the Bill and the Scottish representatives negotiating for the Netherbow's survival well represents the Netherbow's political and symbolic role. Interestingly, during the uprising there was never an argument about whether the king's troops, cut off in Canongate, should forcibly violate the city's urban perimeter. One of the witnesses remarked during the investigations how "had the troops forced their way into the town by demolishing one of the gates, without a legal authority, your grace would soon have had a terrible complaint from the magistrates" (Parliament, Great Britain, 1812, p. 193). The militia's unwillingness to destroy the Netherbow suggests some awareness of the symbolic significance of such an act: achieving possession of and altering one or more civic spaces meant appropriating the narrative that proclaimed the city's past. However, after the incident, razing the Netherbow became a key part of the punishment, and was reminiscent of the traditional breaking down of gates, walls, and other defences as a highly symbolic punishment inflicted on cities that had rebelled against their rightful rulers. In the case of Edinburgh, "taking down the gates of the Nether-bow Port, that a communication might be opened between the city and the suburbs, in which the King's troops were quartered" (*The Scots Magazine*, 1802, p. 659) might sound like a sensible solution against a further uprising. The taking of the Netherbow had indeed led to the mob's unlawful and unauthorized control over the city space, and the Parliament's suggestion to retaliate by destroying the Netherbow to avoid a reoccurrence could then have practical foundations. However, the supporters of the Bill pointedly downplayed—or it is possible that they honestly misunderstood—the significance of the removal of the gateway. One of them declared that "the town-guard ... never can be reckoned among their ancient and immemorial privileges. Nor can I see how either the demolishing of a gate, or the keeping it open, can affect them" (Parliament, Great Britain, 1812, pp. 244–5). However, the Scottish MPs' vehement protestations showed their awareness of the Netherbow's historical and cultural significance (Arnot, 1816, p. 160). Practical reasons were invoked; the financial use of the checkpoint, the defensibility of the road narrowing in general with or without the Netherbow itself, the option to enlarge the army's barracks in the Castle to give the army easy access to the city (Parliament, Great Britain, 1812, p. 240). More importantly, the legitimacy of such an act was questioned. The Duke of Argyle argued that "we may determine the Properties of private Persons, and may adjust the Privileges of Communities, we cannot infringe the Rights of Nations." He also added that "The Nation of Scotland, in all the Proceedings at the Time, treated with England as an independent and free People," and in the Treaty of Union "the Privileges of the Royal Burghs ... were put upon the same Footing with Religion, that is, they were not alterable by any subsequent Parliament of Great Britain" (Parliament, Great Britain, 1812, p. 239). Given Edinburgh's history of royal privileges, the suggestion to lay a proudly walled city bare and unprotected interfered with Edinburgh's sense of self and meddled with rights and prerogatives which had defined Scotland's burghs' identities for 500 years (Anon., 1830, p. 36).

The taking down of the Netherbow holds a much higher significance than a discussion of military accessibility or economic inconvenience. The aftermath of the 1707 Union saw the controversial dissolution of the Scottish Parliament as an independent entity, and the continuous threat to the Hanoverians' rule posed by Jacobean uprisings in Scotland supporting Stuart pretenders. The demolition of the Netherbow became then a point of contention between the Scots' threatened sense of identity and self–determination and English expectations of conformity and uniformity across a newly unified Great Britain. It is quite telling that the amended version of the Bill which was finally approved, imposing a hefty fine on the city but sparing the Netherbow, was disparagingly considered inadequate and stingless by those who had supported its earlier version (Parliament, Great Britain, 1812, p. 190).

CONCLUSIONS

In the early modern period, the Netherbow represented the most visible and commodious entrance into Edinburgh. The city's complex role as a royal burgh, the Stewarts' capital, and a seat of national government, meant that the Netherbow was crossed, attacked, or defended by a variety of people in many different circumstances. Throughout its history, its importance was recognized as a permeable entry and an impermeable defence, but also as a symbol of traditional identity, embodying one of the burgh's fundamental privileges, to think for itself and freely distinguish friend from foe.

Its martial and imposing appearance long survived its effective military usefulness, which was severely challenged by the introduction of artillery warfare. During the siege of 1745, the inhabitants of Edinburgh fled through the Netherbow when the use of batteries and field pieces caused severe destruction to the city, their flight representing their disbelief in the capacity of the defenses to protect them (Grant, 1880–83, II, p. 331). Nevertheless, the Scottish castellated architectural style had been recognized as an intentional reminder of the country's antiquity and martial past, and the significance of the Netherbow went well beyond its military usefulness (McKean, 2004). As late as 1737, the Parliamentary enquiry into the Porteous incident demonstrated the symbolic role attributed to the gateway in defining the city's identity. However, at the end of the eighteenth century new values of productivity and social and economic improvements changed the organization of Scottish society, challenging the acknowledged idea of urban community that the Netherbow represented. The Netherbow's role as urban border was particularly questioned by the construction of a regularized and modern New Town, beginning in 1765, facing the overcrowded and old-fashioned Old Town from the other side of the North Loch, and freely accessible over the commodious and newly built South Bridge. Left behind in the isolation of the Old Town backwater, the Netherbow was finally demolished in 1764 as an obstacle to traffic and circulation, and to allow an easier passage to and from Canongate for both people and carriages (Arnot, 1816, p. 183). The increase in the number of wheeled carriages made the Netherbow "exceedingly incommodious" (Kay, 1842, II/I, p. 186), a problematic relic of a past in which circulation was filtered and controlled. This is a surprising development, coming just 30 years after Argyle's passionate defense of the Netherbow, but it also demonstrates Edinburgh's changed perception of its urban identity. London itself obtained in 1760 an Act of Parliament to remove its own city gates, and the ambitions to compete with London in modernity and stylishness made Edinburgh turn to new constructions embodying the principles of the Enlightenment, such as the New Town, rather than continuing to take more traditional stances (Wilson, 1848, I, p. 114. Youngson, 1966). The Netherbow was not the only victim of the expansion and renovation undertaken in this period, as other civic landmarks were also destroyed, such as the Tolbooth, taken down in 1817, and the Butter Tron, razed in 1822. These improvements were not unanimously welcomed: Scottish poet and satirist Claudero (whose real name was James Wilson) lamented the loss of the Cross in one of his witty works called *"the last Speech and Dying Words of the Cross, which was hanged, drawn and quartered on Monday the 15th of March 1756, for the horrid crime of being an encumbrance to the street"* (Wilson, 1848, II, pp. 219–20).

In a similar work addressing the lamentable destruction of the Netherbow, Claudero reasoned that with its loss, the relationships between people and places were altered, creating confusion in the very fabric of society. In the place of familiar activities, busy workshops and slow-paced, friendly neighborhoods historically connected to and justified by the Netherbow's very presence, Claudero argues that now there is the speed and also the impersonality of modernity, represented by luxurious gilded chariots traveling at a dangerous speed, uninterested in and unresponsive to the surrounding urban spaces and their alienated inhabitants (Wilson, 1848, II, p. 220). In Claudero's poem the city gate is humorously but also sympathetically personified while he presents to the audience its last speech and dying words before demolition. The condemned port recalls many events of urban history not as an impassive observer, but as an involved and passionate actor lamenting the execution of brave Marquis of Montrose and the public display of his severed head, fondly remembering the arrival in town of King James VI, and criticizing Cromwell the usurper (Wilson, 1848, II, p. 221). As wittily illustrated by Claudero, before it was superseded by the unstoppable waves of modernity the Netherbow represented an active agent of urban history, influencing people's understanding of the city and their own perception of self in relation to the urban spaces they moved within.

With reference to the site where the Netherbow stood, not much is left now to commemorate its presence and the major role it had in portraying and defending the burgh's identity. At the narrowing of the street between Edinburgh and Canongate, brass bricks on the street pavement form the contour of the disappeared gate, while commemorative plates and fragments of cornices and pediments are visible in the nearby buildings. The name of a pub at the very crossroad, the World's End, commemorates the vanished frontier between two worlds and it recalls the World's End Close, "the last alley … that terminated of old the boundaries of the walled capital, and separated it from its courtly rival, the Burgh of Canongate" (Wilson, 1848, II, p. 55). Anxious not to be forgotten, Claudero's Netherbow wished in his last speech that his clock be moved to a different location rather than destroyed, so that it would forever look down to "the brandy shops below, where large greybeards shall appear to him no bigger than mutchkin bottle ... reflecting the rays of the sun to the eyes of ages unborn" (Wilson, 1848, II, p. 221). And so, in the spirit of Claudero, perhaps the fact that his role as urban frontier is perpetuated in the name of a public house would provide the old Netherbow with a reason to be pleased.

REFERENCES

Adams, N. 1972. *Dead and Buried? The Horrible History of Bodysnatching*. Aberdeen: Impulse Publications Ltd.

Allen, A.M. 2011. Conquering the Suburbs: Politics and Work in Early Modern Edinburgh. *Journal of Urban History*, 37(3), pp. 423–43.

Alpar Atun, R. and Doratli N. 2009. Walls in cities: A conceptual approach to the walls of Nicosia. *Geopolitics*, 14(1), pp. 108–34.

Anon. 1633. *The Entertainment of the high and mighty monarch Charles King of Great Britaine, France, and Ireland, into his auncient and royall City of Edinburgh*. Edinburgh: John Wreittoun.

Anon. 1822. *Reception at Edinburgh of the Kings and Queens of Scotland, A.D.MDLXI–A.D.MDCL*. Edinburgh: Michael Anderson.

Anon. 1830. *A New Guide to Edinburgh*. Edinburgh: John Anderson.

Arnade, P. 2013. Carthage or Jerusalem? Princely Violence and the Spatial Transformation of the Medieval into the Early Modern City. *Journal of Urban History*, 39(4), pp. 726–48.

Arnot, H. 1816. *The History of Edinburgh, from the Earliest Accounts, to the Year 1780*. Edinburgh: Thomas Turnbull.

Bernheimer, R. 1952. *Wild Men in the Middle Ages, a Study in Art, Sentiment, and Demonology*. Cambridge: Harvard University Press.

Bourdieu, P. 1977. *Outline of a Theory of Practice*. Translated by R. Nice, 1972. Cambridge, MA: Cambridge University Press.

Calderwood, D. 1844. *The History of the Kirk of Scotland*, Vol. V, Thomson, T. ed. Edinburgh: Wodrow Society.

Calderwood, D. 1849. *The History of the Kirk of Scotland*, Vol. VIII, Thomson, T. ed. Edinburgh: Wodrow Society.

Chambers, R. 1836. *The Gazetteer of Scotland*. Edinburgh: Balfour and Jack.

Chambers, R. 1858. *Domestic Annals of Scotland from the Reformation to the Revolution*, Vol. II. Edinburgh: W. & B. Chambers.

Chandler, S. 1679. *The Old Whig; or, the Consistent Protestant*, Vol. II. London: W. Wilkins et al.

Colville, J. and Thomson T. 1825. *The Historie and Life of King James the Sext*. Edinburgh: J. Ballantyne.

Craig, J.T.G. (ed.) 1828. *Papers Relative to the Marriage of King James VI of Scotland*. Edinburgh: Bannatyne Club.

Cullen, W.D. 1988. *The Walls of Edinburgh*. Edinburgh: The Cockburn Association.

Dennison, P.E. 2005. *Holyrood and Canongate, a Thousand Years of History*. Edinburgh: Birlinn.

Dennison, P.E. and Lynch, M. 2005. Crown, Capital, and Metropolis Edinburgh and Canongate: The Rise of a Capital and an Urban Court. *Journal of Urban History*, 32(1), pp. 22–43.

Du Certeau, M. 1988. *The Practice of Everyday Life*. Translated by S. Rendall. Berkeley: The University of California Press.

Foggie, J.P. 2003. *Renaissance Religion in Urban Scotland, the Dominican Order 1450–1560*. Leiden & Boston: Brill.

Foucault, M. 1980. *Power/Knowledge: Selected Interviews and Other Writings, 1972–1977*. Edited and translated by C. Gordon, et al., New York: Pantheon.

Grant, J. 1880–83. *Cassell's Old and New Edinburgh*, Vol. I–VI. London, Paris & New York: Cassell, Petter, Galpin & Co.

Guidicini, G. 2011. Municipal perspective, royal expectations, and the use of public space: The case of the West Port, Edinburgh 1503–1633. *Architectural Heritage*, Vol. 22, November, pp. 37–52.

Haddington, T.H. 1837. *State Papers and Miscellaneous Correspondence of Thomas, Earl of Melros*, Vol. II. Edinburgh: Abbotsford Club.

Harris, S. 1996. *The Place Names of Edinburgh: Their Origins and History*. Edinburgh: Gordon Wright.

Henderson, T.F. 1905. *Mary Queen of Scots*, Vol. II. London: Hutchinson.

Hirschbiegel, J. and Zeilinger G. 2009. Urban space divided? The encounter of civic and courtly spheres in late-medieval towns. In A. Classen (ed.) 2009. *Urban Space in the Middle Ages and the Early Modern Age*. Berlin and New York: De Gruyter, pp. 481–504.

Historical Manuscripts Commission. 1894. *The Manuscripts of Lord Kenyon, Fourteenth Report, Appendix, Part IV*. W.J. Hardy (ed.). London: Eyre and Spottiswoode.

Isidore of Seville. 1911. *Etymologiae*, W.M. Lindsay (ed.). Oxford: Oxford University Press.

Jost, J.E. 2009. Urban and liminal space in Chaucer's Knight's Tale: Perilous or protective? In A. Classen (ed.) 2009. *Urban Space in the Middle Ages and the Early Modern Age*. Berlin and New York: De Gruyter, pp. 373–94.

Kay, J. 1842. *A Series of Original Portraits and Caricature Etchings*. Vol. 1, Part II. Edinburgh: Hugh Paton.

Keith, T. 1913. Trading privileges of the royal burghs of Scotland. *English Historical Review*, 28/111, pp. 454–71.

Knox, J. 1831. *The History of the Reformation of Religion in Scotland*. W. McGavin (ed.). Glasgow: Blackie, Fullarton, & Co.

Kostof, S. 1991. *The City Shaped: Urban Patterns and Meaning through History*. Boston: Little Brown.

Kostof, S. 1992. *The City Assembled: The Elements of Urban Form through History*. Boston: Little Brown.

Kusch, M. 1978. The river and the garden: Basic spatial models in Candide and La Nouvelle Heloise. *Eighteenth Century Studies*, 12(1) Autumn, pp. 1–15.

Laing, M. 1819. *The History of Scotland*, Vol. 2. London: J. Mawman.

Lefebvre, H. 1991. *The Production of Space*. Translated by D. Nicholson-Smith. Oxford: Blackwell.

Liddiart, R. 2005. *Castles in Context, Power, Symbolism and Landscape, 1066–1500*. Macclesfield: Windgather Press.

Lynch, M. 1981. *Edinburgh and the Reformation*. Edinburgh: John Donald Publishers.

M'Gavin, W. 1828. *Biographia Scoticana*. Glasgow: W.R. M'Phun Publisher.

MacDonald, A.A. 1991. Mary Stewart's entry to Edinburgh: an ambiguous triumph. *The Innes Review*, 42(2), Autumn, pp. 101–10.

Mackay, M.W. 1949. *The Scottish Burghs*. Edinburgh: Oliver and Boyd.

Mackillop, A. 2011. Confrontation, negotiation and accommodation: Garrisoning the burgh in post-union Scotland. *Journal of Early Modern History*, 15(1), pp. 159–83.

Mann, A.J. 2010. House rules: Parliamentary procedure. In K.M. Brown and A.R. MacDonald (ed.) 2010. *The History of the Scottish Parliament, Parliament in Context 1235–1707*, Vol. III. Edinburgh: Edinburgh University Press, pp. 122–56.

Marwick, J.D. (ed.) 1882. *Extracts from the Records of the Burgh of Edinburgh 1573–1589*, Vol. IV. Edinburgh: Scottish Burgh Records Society.

McKean, C. 2004. *The Scottish Chateau: The Country House of Renaissance Scotland*. Stroud: Sutton, 2004.

McLynn, F. 2003. *Bonnie Prince Charlie: Charles Edward Stuart*. London: Pimlico.

McManus, C. 2002. *Women on the Renaissance stage: Anna of Denmark and Female Masquing in the Stuart Court 1590–1619*. Manchester: Manchester University Press.

Mignet, F.A.M.A. 1851. *The History of Mary Queen of Scots*, Vol. I. London: Bentley.

Mill, A.J., 1927. *Mediaeval Plays in Scotland*. Edinburgh: William Blackwood & Sons.

Neilson, W.A. (trans.) 1999. *Sir Gawain and the Green Knight*. Cambridge, Ontario: In Parentheses Publications.

Parliament, Great Britain. 1812. *The Parliamentary History of England from the Earliest Period to the Year 1803. Vol. X, A.D. 1737–1739*. London: T.C. Hansard.

Pepper, S. 2000. Siege law, siege ritual, and the symbolism of city walls in Renaissance Europe. In J.D. Tracy (ed.) 2000. *City Walls, the Urban Enceinte in Global Perspective*. Cambridge: Cambridge University Press, pp. 282–316.

Reyerson, K.L. 2000. Medieval walled space: Urban development vs. defense. In J.D. Tracy (ed.) 2000. *City Walls, the Urban Enceinte in Global Perspective*. Cambridge: Cambridge University Press, pp. 88–116.

Rothauser, B.C.L. 2009. The use of water in the medieval consideration of urban spaces. In A. Classen (ed.) 2009. *Urban Space in the Middle Ages and the Early Modern Age*. Berlin and New York: De Gruyter, pp. 245–72.

Scott, A.G. 1833. *A Diurnal of Remarkable Occurents that have Passed within the Country of Scotland since the Death of King James the Fourth till the Year MDLXXV*, Vol. I. Edinburgh: s.n.

Scott, W. 1841. *The History of Scotland*, Vol. II. Edinburgh: Robert Cadell.

Sennett, R. 2004. The city as an open system. In *Leverhulme International Symposium 2004, the Resurgent City, London 20 April 2004*. London: London School of Economics.

Stalder, L. 2009. Turning architecture inside out: Revolving doors and other threshold devices. *Journal of Design History*, 22(1), pp. 69–77.

Stewart, L.A.M., 2006. *Urban Politics and the British Civil Wars: Edinburgh, 1617–53*. Leiden: Brill.

Strong, R. 1984. *Art and Power: Renaissance Festivals 1450–1650*. Woodbridge: Boydell and Brewer.

Testaverde, A.M. 1980. Feste Medicee: la Visita, le Nozze e il Trionfo. In M. Fagiolo Dall'Arco (ed.) *La città effimera e l'universo artificiale del giardino*. Roma: Officina Edizioni.

The Scots Magazine. 1755. Vol. 17. Edinburgh: Sands, Donaldson, Murray, and Cochran.

The Scots Magazine. 1802. Vol. 64. Edinburgh: Alex Chapman and Co.

Tracy, J.D. 2000. To Wall or not to Wall: Evidence from Medieval Germany. In J.D. Tracy (ed.) 2000. *City Walls, the Urban Enceinte in Global Perspective*. Cambridge: Cambridge University Press, pp. 71–87.

Turner Simpson, A., Stevenson, S., and Holmes, N. 1981. *Historic Edinburgh, Canongate & Leith, the Archaeological Implications of Development*. Glasgow: Scottish Burgh Survey.

Weber, M. 1958. *The City*, first published 1921. Translated by D. Martindale and G. Neuwirth (eds). New York: Free Press.

Wilson, D. 1848. *Memorials of Edinburgh in the Olden Time*, Vol. I–II. Edinburgh: Hugh Paton.

Youngson, A.J., 1966. *The Making of Classical Edinburgh 1750–1840*. Edinburgh: Edinburgh University Press.

PART II
Border Buildings

5

House Number 1: The Vienna Hofburg's Multiple Borders

Richard Kurdiovsky

Royal, princely, imperial or ducal palaces are usually hermetically sealed off from their surroundings for reasons of security and to emphasize their character as the seat of power and of the sovereign. The instruments to visualize this special character can be topographical, architectural, and stylistic, among others. Buckingham Palace, for instance, remains aloof from the lively metropolis of London by means of large parks and green areas, broad avenues where public traffic can be halted if necessary, and strong fences to keep people at a respectful distance. In addition to national borders and city limits, frontiers can therefore be found within cities and societies, dividing spaces of elites from those for people of subordinate status. The visual appearance of this special type of border is quite different from that of national borders. While at first sight these borders may look well-established (facades of buildings, fenced-off areas, etc.), they can turn out to be negligible in certain situations and to be characterized by overlap or permeability under special circumstances. In this sense, borders are not merely touchable barriers, but social and functional barriers within a city that can physically be crossed easily, perhaps unawares, and yet they can just as easily be completely closed. The following essay takes the Hofburg in Vienna as an example of such a border and concentrates on its nineteenth-century state until the end of the Austro-Hungarian monarchy in 1918.

The Vienna Hofburg (Dreger 1914; Lhotsky 1941; ÖZKD 51 [1997]; Rosenauer 2012ff) was the official residence of the Habsburg emperors of the Holy Roman Empire (until 1806) and of the emperors of Austria (from 1804 until 1918) who were at the same time kings of Hungary and Bohemia with residences as well in the castles of Buda and Prague. The Hofburg in Vienna served as the emperor's and his family's personal home in wintertime (usually from late autumn until spring) with all the necessary facilities to meet a great household's needs. It also served as the official seat of power of the Habsburgs as monarchs, with all necessary court offices including political and military institutions and, not least, storage facilities for the insignia of their reign. The Hofburg was the place where the majority of family-owned goods were kept, such as books in the court library, art objects in various rooms and galleries, natural scientific collections in different cabinets, jewellery and other precious items in the treasury, and so on. Cultural institutions such as the court theatres (consisting of the theatre and the opera in separate buildings from the eighteenth century onwards) also formed part of the Hofburg. Thus, we must imagine the Hofburg as a multi-functional complex that not only met the demands of the monarch, of his aristocratic entourage, and of their subordinates and servants, but that also served the interests of a wider public, including people who did not belong to court society at all. Only in the final years of the Habsburg monarchy, however, did the Hofburg reach an architectural state that could be distinguished more or less clearly from its non-courtly surroundings. Indeed, to ask about the Hofburg's borders and how one could distinguish it from the city around it turns out to be a rather complex question.

THE TOPOGRAPHICAL SITUATION OF THE HOFBURG

Around the middle of the thirteenth century, the oldest part of the Hofburg (see Figure 5.1) was built along the city walls as a four-winged fortress with corner towers forming the so-called Schweizerhof (Swiss Guards' Court), which remains today the heart of the whole Hofburg complex. Towards the city to the east, there ran one of the most important traffic routes of Vienna, the Herrengasse (Noblemen's Street) and the Augustinerstraße (Augustinians' Street) that together connected the main city gates: the Schottentor (Scots Monks' Gate) in the north and the Kärntnertor (Carinthian Gate) in the south. Where this street met the area of the Hofburg (in the area of today's St Michael's Square), there was an intersection with another artery, the Kohlmarkt (Coal Market). This street led to the only city gate that gave access to the city from the west, and the medieval Hofburg was located right next to it. Therefore one of the main traffic routes into the city passed by what later became the central part of the Hofburg.

In the course of the sixteenth and seventeenth centuries, the medieval Hofburg was enlarged with individual buildings that were connected to it via galleries and separate wings. These buildings stretched out over the civic neighborhood, incorporating once privately-owned properties into the area of the Hofburg. Thus streets and squares that were once areas of public traffic (such as the above mentioned route through the city gate) were turned into courtyards and passages. At certain points, bridges crossed over the narrow streets around the Hofburg to connect single lots to the palace complex. Because of this specific architectural development of the Hofburg that reached out from a core building over parts of the surrounding city, the court had to come to an arrangement with the urban public, as will be discussed later.

Around the same time, Vienna's fortification system (see Figure 5.2) was modernized according to early modern standards and new ramparts and bastions

5.1 Bird's eye view of the Hofburg complex looking east, photograph taken between 1913 and 1931. Reproduced courtesy of ÖNB, Pictures, Archive and Graphic Department, Inv. No. 39.866-C.

were built (ÖZKD 64 [2010]). Besides their military function, they provided extension areas for the Hofburg that were used, for example, as hanging gardens for the imperial court or as building sites for temporary theatres. Having reached its final dimensions by the end of the seventeenth century, different parts of the Hofburg were altered several times by replacing old buildings with new ones or just by changing the decorations of the facades. But the perimeter of the Hofburg complex remained more or less unchanged.

In the middle of the nineteenth century, the city fortifications were demolished, making way for the Ringstraße boulevard. Here, the most voluminous extension project of the Hofburg, the Kaiserforum (Emperor's Forum), was begun but never completed. Occupying the whole depth of the glacis (that is, the plain in front of the early modern ramparts that was kept free of any building for surveillance and clear artillery firing), new wings should have been added to the Hofburg that enclosed a gigantic inner square. Of these wings, only the Kunsthistorisches (Art History) and Naturhistorisches (Natural History) museums and the wing for the imperial apartments were actually realized. At the far end, the unfinished Kaiserforum still ends in the extensive baroque facade of the imperial stables (Hofstallungen), built in the 1720s at the glacis' edge, which at the same time served as the innermost building line of the suburbs.

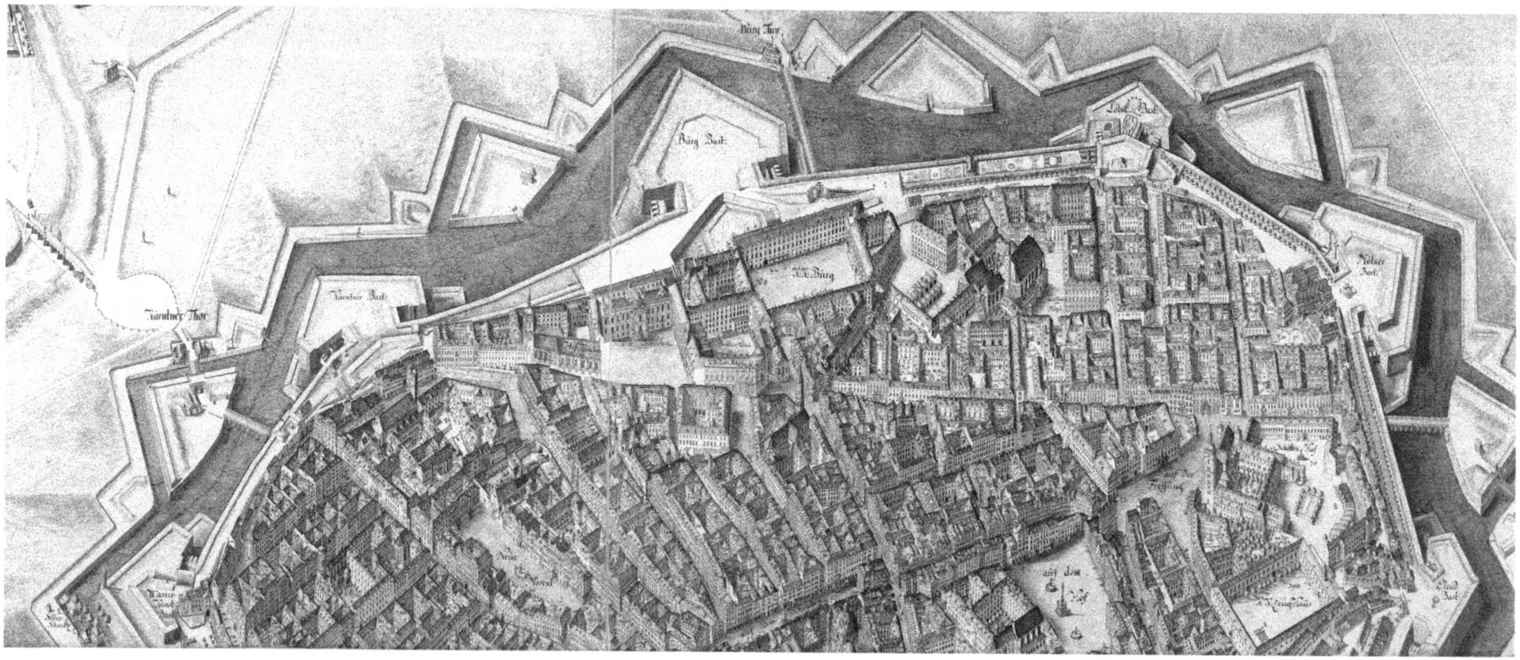

5.2 Bird's eye view of the Hofburg complex looking west, drawing by Joseph Daniel Huber, 1778. Albertina Vienna, Inv. No. 37.074 and 37.075. Reproduced courtesy of Albertina, Wien, www.albertina.at.

MASSIVE BORDERS: FORTIFICATIONS

The city fortification that delimited the Hofburg towards the suburbs never changed its position and its military character, which included massive walls and a small number of gates. Especially after Vienna was unsuccessfully besieged twice by Ottoman troops in 1529 and 1683, these city walls became a symbol of Christian Europe's stronghold against the heathen Turks, gaining military and political prestige for the ruling Habsburg dynasty at the same time (Feichtinger 2010). Although their military functionality had already been questioned in the course of the eighteenth century (Mader-Kratky 2015), the city walls remained, until their demolition in the 1850s, a clearly readable sign separating the city of Vienna from its suburbs (with the glacis as an extensive buffer zone). This distinction was not only an architectural but also a social one: within the city walls, there was the main residential area of the nobility and of upper bourgeois circles, while the suburbs mainly served as the living quarters for craftsmen, lower civil servants, and workers who could not afford the high rents of the so-called Innere Stadt (Inner City) (Lichtenberger 1978; Weigl 2003, Schwarz 2010, esp. 132–3). Physically, both the city of Vienna and the Hofburg shared the highly symbolic element of the city walls due to the Hofburg's position along these walls.

But the Hofburg was the only building of a prominent size that overlooked the city walls and had direct access to the ramparts. Responsibility, both military and administrative, as well as access were not shared between the Hofburg and the city. The city walls, as all the other parts of the defensive system of Vienna, were military areas with rigorous restriction of access. All civic building was forbidden there, and only members of the guard troops were allowed to erect small huts as shelters and primitive living quarters. On the other hand, the huge flat surfaces on top of these walls offered an ideal site for laying out, for instance, baroque gardens—but again just for the use of the emperor and his family, and thus explicitly expressing the supremacy of the sovereign over his subjects.

The situation remained unchanged until 1785, when Emperor Joseph II opened the ramparts and bastions for the general public as a resort for promenading and recreation. If we believe contemporary descriptions, the whole social stratum was actually able to participate (Eberle 1911; Till 1959), although the cafes that were opened in former court pavilions on these premises mainly attracted wealthy members of Vienna's society. The city walls, therefore, became a zone of overlapping interests (albeit it did not become a zone of intermingling contact) between the court and the civic public. But the city walls, as all other parts of the former defensive system, remained state property and the court bore administrative responsibility for them. Already from this first example of the fortifications, we learn that visual borders could be shared by the Hofburg and the city. And at the same time, they could be a clear sign of social and administrative separation and could indicate the borders of influence and political power. Questions of use of and responsibility for these borders could be dictated by the head of state and his officials, even if they were meant for the common welfare.

STREETS AND SQUARES AS BORDERS OF THE HOFBURG

In general, streets clearly delineate the borders of buildings or of groups of buildings in a city. Unfortunately, this does not apply to the Hofburg. Although it was (and still is) encircled by a sequence of streets, this does not mean that, if you walked around the complex in a clockwise direction, whatever belongs to the Hofburg would be on the right hand side. Buildings that did not belong to the Hofburg "intruded" into its perimeter. Among these were a number of private bourgeois houses and noble properties in the area of today's St Michael Square. These parcels could not be bought up by the court until the 1890s, and this blocked the completion of the Hofburg's facade facing this square. Another example was the Augustinian monastery: its huge church was not only used very often for court religious ceremonies, but the Habsburgs spent substantial sums of money for the decoration and remodeling of the church and its chapels. Still, the monastery itself remained an independent unit within the Hofburg complex. A third example was the Albertina palace: in contrast to the above-mentioned private houses, it actually formed part of the Hofburg and housed different court offices until 1794, when it became the private property of Archduke Albert of Saxe-Teschen (Benedik 2008, 35). Even after it was decorated with new facades that clearly separated it visually from the rest of the Hofburg, it remained in fact physically attached to the imperial palace.

In contrast to these "intruding" private properties, the Hofburg itself could "intrude" into its neighborhood: bridge constructions crossed the streets around the Hofburg twice to connect the palace with satellite buildings that could either be attached houses or could form an independent architectural entity. One of these, the Hof- or Kaiserspital (Court or Emperor's Hospital), demolished in 1903, formed part of a block of private houses. Only its facades were visible from the street, while the fire-walls were shared with the neighboring houses. In these cases, the actual course of the border between the court's sphere and the civil environment could not be assessed with precision by passers-by; it was only visible on abstract media such as maps. The other "intruding" building, the Stallburg (Stable Palace), remains a free-standing rectangular building following the building type of a cubic-shaped renaissance palace. Since it thus forms an independent piece of architecture, it seems no more than marginally attached to the Hofburg by means of a bridged arch. When the Stallburg was built around the middle of the sixteenth century, its appearance closely resembled that of other facades of the Hofburg in shape, decoration and the colour of the building material. That all these buildings belonged to one and the same complex was much more evident once that it looks today (Holzschuh-Hofer 2010).

But the Hofburg did not only protrude into the surrounding city, it also allowed "public" space to extend into court territory. In early modern times, the area of today's Josephsplatz (a square named after emperor Joseph II) served as gardens and as an open riding school for the exclusive use of the imperial court. Surrounded on three sides by court buildings and other components of the Hofburg, it was screened from nearby Augustinerstraße on the fourth side by a massive wall (called

Augustinergang, Augustinian Passage or corridor). Whatever was behind this wall was hermetically sealed off from public view—as for instance the famous baroque court library building by Johann Bernhard Fischer von Erlach, built at the opposite end of this inner open space. The Augustinergang was not torn down until the mid-eighteenth century, providing, for the first time, a view onto the magnificent library building and creating, again for the first time, a square with facades built from 1769 until 1776 (Benedik 1993, 14–16) that equaled the Places Royales in France with their unified facades. Since there was no fence, no haha-ditch or any other sign of a demarcation between the public street and what once formed part of the Hofburg, visible borders between the court's sphere and the zone of the urban public did not exist. The ground area nevertheless remained court property. Therefore, court officials could allow or refuse requests handed in by the municipal authorities to use the Josephsplatz for instance as a temporary market place for wood and ironware (Austrian State Archives [ÖStA], House-, Court- and State Archive [HHStA], court board of work [HBA], 304 and 884 ex 1792). The encoding as an imperial space, nonetheless, remained clearly legible, not just by means of the impressive facades, but also because of the equestrian statue of the emperor Joseph II, erected in 1807, that helped to reinforce the presence and dominance of the court and the state.

Borders of palatial structures, therefore, need not necessarily follow a clearly visible course that separates the palace from its surroundings. Depending on functions and the specific history of a site, these border areas can also be intertwined with each other.

FACADES AS AMBIVALENT BORDERS

Another significant feature that represents the border between the Hofburg and the city emerges from these considerations: the Hofburg's facades. A facade can be understood as the skin or clothing of a building that is shown to the public and that carries all the signs and symbols that indicate how it is to be seen by others. A characteristic example is the facade towards St Michael's Square that was realized following designs of the baroque architect Joseph Emanuel Fischer von Erlach (see Figure 5.3). A photograph taken around 1895/1896 (Figure 5.4) gives a specific view on the urban situation and appearance of the imperial palace complex in relation to the civil urban structures. The choice of architectural vocabulary strikingly differentiates the imperial sphere from its neighborhood. At first sight, we notice the difference in height—the Hofburg facade on the left reaches nearly twice the height of the Palais Dietrichstein on the right—which makes the Hofburg appear more voluminous, as if it occupies a greater amount of space. Accordingly, proportions seem enlarged; the size of the windows and the height of the rusticated basement have been increased. The outline of the Hofburg with its cupolas is much more elaborate than the simple saddleback roof of the Palais Dietrichstein. The use of columns for the piano nobile zone and of rich sculptural decoration on the attic balustrade further sets the imperial architecture apart from its surroundings. Although the Palais Dietrichstein itself was the town residence of a noble family, it

5.3 Joseph Emanuel Fischer von Erlach, design for the completion of the Hofburg-facade towards St Michael's square. Engraving by I.A. Corvinus after a drawing by Salomon Kleiner, published in 1733; Austrian Academy of Sciences, collection Woldan. Reproduced with permission of the Osterreichische Akedemie der Wissenschaften, BAS:IS.

was obviously dominated by the palace of the monarch in an architectural sense. The same could be said about the other end of the Michaeler facade opposite the Kleines Michaelerhaus (Small St Michael's House), an apartment block for bourgeois families. Here even an additional floor did not guarantee that the building would reach the same total height as the Hofburg.

But this socially hierarchical differentiation was clearly legible only for a short period of time: from the completion of the Hofburg's facade towards St Michael's Square in 1895 until the demolition of the adjacent Palais Dietrichstein in 1897. The new building that was erected on this site used the same means and vocabulary, thus equaling the Hofburg in size and in the use of columns and sculpture. Harsh criticism followed immediately (Tietz 1910, 52): With his design, the architect crossed the border of social decorum because he had used prestigious elements that, according to French eighteenth-century architectural theory, should be reserved for princely palaces only. What people could observe until construction on the facade towards St Michael's Square started in 1890 was completely different (Figure 5.2), and actually consisted only of fragments of the baroque project of Joseph Emanuel Fischer von Erlach that should have replaced the old entrance to the Hofburg (a narrow street that ran between the courtly Ballhaus, a hall for ball games, and court chancelleries and private houses) with an impressive new main facade towards the city (Figure 5.3). But to do this, it would have been necessary to tear down a number of court buildings as well as privately owned houses. In 1726, the project was begun under the reign of Charles VI on those properties that

5.4 Photograph of the northern end of the Hofburg-façade towards St Michael's square and the adjacent Dietrichstein-palace prior to its demolition in 1897. Reproduced courtesy of ÖNB, Pictures, Archive and Graphic Department, Inv. No. Cl.203.10-10A.

already belonged to the court (the left-hand parts of the facade and the central rotunda). It was never completed, however, because buying the private properties turned out to be impossible, thus leaving the project in a half-finished state, with some walls unplastered. From 1741 onwards, the whole plan was changed. The old Ballhaus was not demolished, but turned into a theatre: the old Burgtheater. Its new facade, built in the 1760s, followed a completely different model and showed no relation to the facade scheme that was started some decades earlier. The private houses were left untouched, and therefore the imperial Hofburg was hidden from sight by a row of civic buildings that followed the model of Viennese bourgeois houses (Figure 5.2). Seen from Kohlmarkt, one could hardly perceive the Hofburg as a piece of palatial architecture. In contrast to the fortifications towards the suburbs, the facades of the Hofburg towards the city center did not indicate exactly where the sphere of the imperial court began—at least not until the last decades of the Habsburg monarchy.

WAYS TO ENTER THE HOFBURG

Neither was the way to enter the palace of the emperors of the Holy Roman Empire as dignified as one might imagine. Descriptions from the early eighteenth century already point towards the unsightly appearance of the imperial palace (e.g. Freschot 1705, 3–6). But even then, the moment when one would have crossed the border of the Hofburg could be recognized by some by taking an element that was both architecturally as well as functionally insignificant as the symbol of the imperial threshold. When a feudal ceremony took place in 1853 (ÖStA, HHStA, ceremonial protocols [ZA-Prot.], 65 ex 1853), the prince-archbishop of Olomouc (Moravia, Czech Republic) and his entourage used the traditional route of approaching the Hofburg by way of Kohlmarkt and St Michael's Square. In the open, unfinished rotunda, they stopped under the "Theaterschwibbogen" (one of the rotunda's arches next to the Burgtheater). In the ceremonial protocols, this arch was described "as the entrance to the Hofburg" ("als dem Eingange in die Burg"). The cortege (explicitly the servants) even had to take off their hats as if they had already entered an interior space. Unlike Marble Arch, the main entrance to Buckingham Palace until the mid nineteenth century, it was not a triumphal arch that marked the point of access to the Hofburg. Actually, it was no portal at all but a simple and minor archway where theatregoers would usually queue before the rise of the curtain. Clearly, it could be interpreted as the official entrance to the Hofburg although nobody would have taken it as such based simply on its appearance and situation.

Until the 1890s, there were hardly any distinctive architectural signs accentuating the border between the imperial court and the general public that could clearly be read and understood: when approaching the Hofburg from Kohlmarkt, one would have passed through a building complex that looked like any urban district with houses of different sizes and shapes. In contrast, when the Hofburg's facade towards St Michael's Square was completed in 1895, using the street to the west meant that one passed through the impressive portal of a palace. The exclusive and the ordinary spatially merge, since everyday traffic could pass through the Hofburg, as will be described later on. However, when the archbishop's entourage proceeded towards the Schweizerhof and the state rooms, they had already crossed the border of the Hofburg and entered the sphere of the imperial court's aura; at the same time and only some meters next to them, other people could be in the same area without having crossed this border. One and the same space can be interpreted as part of an exclusive sphere (the courtly sphere of, for instance, imperial receptions) or as a public space of everyday life. Unfortunately, the sources from the court archives do not tell us whether these two situations did occur in practice at one and the same time, or whether public traffic was interrupted during such court ceremonies.

STREETS, TRAFFIC AND TRANSPORT THROUGH THE HOFBURG

This leads us to another point closely connected with streets that shows the Hofburg's elements of functional and social overlap. Aside from marking borders, there is another quality of streets that is important for the Hofburg: streets as routes for traffic and transportation (Telesko, Kurdiovsky and Sachsenhofer 2013, 40–47). As mentioned above, the only city exit to the west was situated close to the medieval Schweizerhof. By the late seventeenth century, the square in front of the Schweizerhof was completely surrounded by court buildings and had been turned into an inner courtyard of the Hofburg complex; in the nineteenth century, this courtyard was called, significantly, Innerer Burghof (Inner Hofburg Square). Concerning the Habsburg court ceremonial, this courtyard was of greatest importance because it served as the main forecourt: from here any visitor of rank (such as ambassadors) would pass through the main gate of the Schweizerhof and enter this innermost courtyard of the Hofburg. The actual city gate, now called Inneres Burgtor (Inner Hofburg Gate) was built into the Lepopoldine wing or Regierender Trakt (Governing or Ruling Wing) that closed off the Inner Hofburg Square towards the city fortification. As indicated by its name, this wing housed the state rooms and imperial apartments that looked onto the Inner Hofburg Square but were accessible only by means of the Schweizerhof. There were no portals towards the Inner Hofburg Square that would have given direct access to the imperial rooms in the Leopoldine wing, only minor entrances for guards and service use, thus indicating that the square in front of the Regierender Trakt belonged to the outside sphere of the Hofburg. Since the city gate was architecturally incorporated into this wing, any traffic of people as well as any transportation of goods that moved from the city towards the city gate automatically crossed the Hofburg. It was even "penetrated," since public traffic with all its noise and dust had to pass through the Regierender Trakt on ground floor level, just two floors beneath the state rooms! Imperial and court spaces were not left untouched and could not be passed unnoticed. In this context, a description by Franz Xaver Schweickhardt of Sickingen of the sound-proof windows in the apartment of emperor Francis II/I is quite elucidating: "that one does not hear the rattling carts driving through the Hofburg when looking down on the square swarmed by people and carts constantly driving back and forth" ("daß man keinen Wagen in der Burg fahren hört, [...] wenn man auf den Platz hinuntersieht, wie alles von Menschen wimmelt, und viele Wagen beständig hin und wieder fahren [...]", Schweickhardt 1932, 9–17).

This thoroughfare was of interest both for the city and for the court. The city needed this street for public traffic and the court needed it, for instance, as the entrance route ambassadors would take on their way to be accredited at the Habsburg court. Therefore it was in the court's interest to maintain this street in the best possible condition: already in the sixteenth century, the Hofbauamt (the court's board of works) was responsible for paving and cleaning this street (Jeitler and Grün 2014). By the late eighteenth century, the city's interest seems to have prevailed: now, the municipal Unterkammeramt (lower chamber office, i.e. the city's department for building activities, fire protection and other infrastructural demands) was responsible for repairing this street—but just the street and nothing else. When city authorities started to repair the street's paving in 1792, the emperor Francis II explicitly expressed his wish that the area between this street and the facade of the Schweizerhof should be renovated as well but, significantly, at the court's own expense, since this ground area did not serve the public (ÖStA, HHStA, HBA, 616 ex 1792). So there was a clear distinction made as to who was responsible for what: the city had to look after the street, the court had to care for the rest of the ground area. This special kind of border was not so much a physical one as a functional, organizational, and financial one. Again, as on the Josephplatz, there were no fences or barriers to separate the street from the rest of the courtyard. The use of the space was shared between the court and other, subordinate social groups. The border here was a site of an overlap of interests in one and the same space.

To ensure dignified comportment when people crossed the court sphere and to maintain imperial decorum, certain rules of conduct were established. Especially when new means of transport such as automobiles or bicycles became widely accepted, interdictions on riding, pushing, and even carrying a bike across the Hofburg complex were imposed (ÖStA, HHStA, office of the Grand Master of the Household [OMeA], rubric 21/9 ex 1901). Honking, for instance, when crossing the Inner Hofburg Square was explicitly forbidden in 1910 and 1911 (ÖStA, HHStA, OMeA, rub. 21/11 ex 1910 and ex 1911). So behavior (as will be described later on in more detail) that guaranteed the desired decorum also signaled the border. Since the Hofburg was characterized by a certain number of permeable borders that let everyday life flow into the courtly spheres, it was obviously important to notice when one would have crossed the borders of the Hofburg complex, even if or indeed because they were visually non-existent.

STEPPING ACROSS THE BORDERS OF THE HOFBURG: IMPERIAL COLLECTIONS AND THEIR VISITORS

Aside from the open space within and around the Hofburg that was shared with the city of Vienna, the interiors of the imperial palace were subject to questions of borders, too. Those parts of the Hofburg where the collections of art and natural science as well as books were kept especially lend the Hofburg the character of an institution of public education. Other spaces of the Hofburg, such as the large halls, were used for cultural performances and amusements for a public audience. And there were parts of touristic interest, too.

The Hofburg therefore was at the center of attention of the general public. In particular, the imperial collections of art and natural science (today kept in the Kunsthistorisches Museum and the Naturhistorisches Museum) constituted areas that were open to public view. They were regarded primarily as scientific and educational institutions rather than as touristic destinations, especially since they were open more frequently for scientific and scholarly research purposes; the court library, for instance, was open to students, scientists, and scholars of all ages without regard to social status or origin. Not only did restrictions of access to any of these collections not seem to have existed (in early modern times in contrast, we find striking examples of how to exclude lower classes from entering the interiors of the imperial court and of how borders were erected by means of one's financial situation: under emperor Charles VI, a visit to the Kunstkammer [art chamber where precious pieces of art and natural wonders were displayed] cost twelve florins and a visit to the treasury even twenty-five florins. Only wealthy [bourgeois or noble] people could afford to pay such incredibly high sums and even then, groups of six or seven persons gathered to share this special "entrance fee" that was regarded as a "tip" [Lhotsky 1941–1945, 392]). But in the case of the court library, the situation became just the opposite; the number of readers consulting the library constantly increased from the seventeenth century onwards. In the second half of the nineteenth century, public interest had grown to such an extent that any changes that were uncomfortable for readers were criticized in the daily press, and were often annulled once these criticisms had been published (Mrázek and Schwab 1968, 464; Mayerhöfer 1968, 529–30). In the course of time, permission to read a book from the court library became a public right that could be claimed by everybody.

But there were both spatial and temporal means of segregating the court from the visiting public. Those wings that housed the imperial collections were either located at the outer zones of the Hofburg complex or they were accessible only at certain times of the year. During the summer months, as part of what was known as the allerhöchster Séjour (the highest sojourn), the emperor, his family, and his court went to stay at one of the summer palaces, usually at Schönbrunn, leaving the Hofburg deserted. Consequently, the general public was able to visit the treasury (that was located in the Schweizerhof close to the state apartments) on two days a week by prior appointment beginning around the middle of the century (Realis 1846, 162) and on three days a week by 1870 (Bucher and Weiss 1870, 73). Similar regulations applied to the Mineralienkabinett (Minerals Cabinet), the Münz- und Antikenkabinett (Cabinet of Coins and Antiquities) and the Zoologisches Kabinett (Zoological Cabinet) (Realis 1846, 109, 111–12; Bucher and Weiss 1870, 74). Furthermore, touristic demands were met: as early as 1846, "the apartments or inner chambers [...] can be viewed in the absence of their majesties and highnesses" (Realis 1846, 85). During the summer sojourn and after submitting the necessary application, visitors were able to view the imperial state rooms where the most illustrious audiences and most important court festivities took place. Right up to the end of monarchy in 1918, such tours even included the major guest apartments that were reserved for the most important guests of state (Petermann 1925, 141 [point A], 148).

To see collections such as the treasury (Lippmann 1881, 38) or to visit the state apartments during the court's sojourn (Petermann 1925, 141), visitors had to enter the Schweizerhof that was accessible by the Schweizertor, and from there take minor doorways to the sides. But this Schweizertor was not just any gate; it was the main ceremonial entrance to the state rooms. Only the coach of an ambassador was allowed to drive through this gate; the coaches of his entourage had to halt outside in the Inner Hofburg Square in front of the Schweizertor. In the open air of the Schweizerhof, there was the first flight of stairs where court officials received and welcomed the emperor's guests; according to their rank and status, this occurred either at the top or at the bottom of the stairs. In exactly the same space, then, through which tourists passed on their way to (just) another fascinating sight in Vienna, architectural features were used to publicly represent the rank of representatives of foreign states.

Old photographs and watercolor vedutas show guards positioned at the Schweizertor; when stepping across this Hofburg border you would have been watched but not prevented from walking through. For instance, it was obligatory to register at the Burghauptmannschaft (the Stewart's office) if one wanted to visit the state apartments (Lippmann 1881, 38). But this office was accessible only via the Schweizerhof. Therefore guards at the Schweizertor could not really act as an admission control as we would understand it today. The consciousness of having crossed the border of the Hofburg was guaranteed by the conduct of people who, once inside the Hofburg, were under more or less constant surveillance (either by guards, by court officials who would guide people through the state apartments, or by the mere fact of the necessary applications). Thus borders could be understood as behavioral lines of demarcation. But as will be shown in the following example, access was confined to those more or less wealthy circles of society who were capable of consuming, for instance, industrial products.

PUBLIC EVENTS AND PUBLIC PERFORMANCES

In 1835, following the explicit wish of the emperor, the first Gewerbsprodukten-Ausstellung (Exhibition of Industrial Products) took place. It was the first of a series of exhibitions that was meant to stimulate the economic development of Austria's industries, and it was hosted in the winter riding school of the Hofburg. In an approach typical of the paternalistic attitudes of the Austrian Empire before the revolution of 1848, the state not only oversaw the economic development of its industries, but it also organized and hosted such an exhibition in the home of the sovereign. To enable everybody to witness the quality products and excellent performance of Austrian companies, the emperor granted free access and "success with the public was very high" ("Der Zuspruch des besuchenden Publikums war sehr groß [...]" [Bericht s.d., 10]). The winter riding school was chosen as the venue because it was not only the most ample space of the whole Hofburg; it was the largest secular space available in the whole city of Vienna.

This was the reason why even public concerts, balls, or other entertainment activities took place within the borders of the Hofburg. When the Gesellschaft der Musikfreunde (Society of the Friends of Music) was founded in 1812/1813, its first big concerts took place in the winter riding school, since it could host up to approximately 5,000 people. Although these concerts were public performances and would have been open to everybody, the court did not have to fear any untoward incidents if it allowed these concerts to be performed within the Hofburg. The cheapest ticket cost two florins and was therefore affordable only for well-off members of the civic population, another example of the social, financial and economic character the borders of the Hofburg could adopt. Nevertheless, the fact that a public concert that had nothing to do with the court, and that was organized by a private institution completely independent of the court, could take place inside the Hofburg suggests that it must have become habitual to enter the Hofburg not for the purpose of attending court activities, but just to listen to musical performances and for entertainment—the Burgtheater and the opera in the Kärntnertortheater (Carinthian Gate's Theatre) were other court institutions that were frequently visited by a public audience.

Apparently, the Redoutensäle [Redouten halls] hosted a great number of performances and events that were far from being part of the courtly life in the Hofburg. As a result, court officials could easily suggest to expel these halls from the Hofburg completely since, in their eyes, a more appropriate location could be found either in a town-hall or in any other building that was owned by the municipality (ÖStA, HHStA, OMeA, rub. 89/20 ex 1863). There was an obvious consciousness of the inconsistency between the Hofburg's main function as the seat of stately power and its actual use in the public and common interest. And yet, wintertime balls that were not organized as part of the ceremonial court festivities (like the exclusive Hofball [court ball] or the even more elitist Ball bei Hof [ball at court]) continued. In 1880, even the emperor and the empress appeared at the ball of the industrial companies in the Großer Redoutensaal, a most noteworthy honor for a civil ball (ÖStA, HHStA, ZA-Prot. 103 ex 1880: 28 January).

ERSTE GESELLSCHAFT AND PRESENTABILITY: BARELY CROSSABLE SOCIAL BORDERS

The general public of Vienna had a strong interest in crossing the borders of the Hofburg for different reasons, and they were not prevented from entering, crossing, or using the Hofburg. The Hofburg was hardly hermetically sealed off from the city around it. Only on certain occasions were the gates closed to avoid gatherings, disturbances and congestion as, for instance, on June 21, 1880, when the 8th dragoon regiment passed through the Inner Hofburg Square with the emperor present. Even then, closure took place discreetly and at the last possible moment (ÖStA, HHStA, ZA-Prot. 103 ex 1880: 21 June); even at night, the gates of the Hofburg were kept open for public thoroughfare, at least for pedestrians (ÖStA, HHStA, OMeA, rub. 21/B/4 ex 1853).

Nonetheless, the Habsburg court remained clearly separated from the civic public around it. Such separation was maintained not so much by means of monumental architecture or impressive town-planning, as by strict regulations of access to court performances and by social codes of conduct and standards of behavior used by members of the imperial family and their subordinates, regardless of whether they were members of the "hoffähige," i.e. presentable nobility, or were among the court servants. Strict exclusivity helped to preserve the court's social decorum and ensured the Hofburg's character as a distinct space that stood out against the mainly bourgeois city surrounding it—the imperial court and especially the emperor Francis Joseph I obviously had such a strong air of exclusiveness that the main protagonist in Robert Musil's *Mann ohne Eigenschaften* (The Man without Qualities), Ulrich, could even doubt that there ever existed the person of the emperor. This exclusivity, which can be traced back at least to the early eighteenth century (Küchelbecker 1730, 377–81), was characteristic of Viennese court society with its emphasis on social distinction (Schneider 2010). Called Erste Gesellschaft (first society), it actually consisted of members of the presentable nobility only, strictly distinguished from all lower social classes, especially from the Zweite Gesellschaft (second society), the group of ennobled banquiers, industrial magnates and big merchants who were dominant members of Vienna's civic society. But it is again remarkable that these separate social groups could use one and the same space within the Hofburg without losing their distinction. The court ball and the ball at court, festivities to which only a very small number of invariably presentable guests were invited, were held in the same spaces where many other events with general public access occurred.

The growing sense of the public's entitlement to participate in some way in events held in the Hofburg is exemplified by a news account of the Carroussel performed in April 1880 (ÖStA, HHStA, ZA-Prot. 103 ex 1880: 17 April). This was a tournament and equestrian game to benefit the needy in Moravia, Silesia, Gorizia, Istria, and Galicia, performed by members of the Austrian aristocracy dressed in historical costumes. Only a small number of visitors were able to watch this performance from the galleries, since the whole arena was occupied with riders and horses. Therefore, the performance had to be repeated two times. Still, the number of spectators was so small (not even including all presentable members of aristocratic society) that long descriptions of the performances were published in the local newspapers. One article (Anon. 1880, 1) seems particularly instructive.

After describing all the grandeur and splendor of the hall and its decoration, of the performers and their noble ancestry, of their costumes and their artful horsemanship, the author discovers a certain shadow and a deplorable lack of "one big and powerful thing": the absence of the people. Reminiscent of baroque court festivities in the same winter riding school, he tells the reader about a celebration under the reign of Maria Theresa (performed on January 2, 1743). She explicitly wanted, once the festivity was over, the whole company to ride through the streets of Vienna to allow the general public to catch a glimpse of them and to participate in the spectacle representing the glory of the house of Habsburg. In 1879, just a year before the Carroussel in the winter riding school took place, another gigantic festive procession had been arranged by Vienna's artists to celebrate the silver wedding of the imperial couple. The coaches and historical costumes were designed by the famous painter Hans Makart. The procession passed in the open air along the Ringstraße, thus allowing everybody, not only the imperial family and the members of their court, to watch this impressive performance. So the citizens of Vienna were already acquainted with large-scale performances in historical costumes and decorations arranged for the highest court and at the same time performed in front of a public audience. The severe critic of the Carroussel performance of 1880 that was published in the newspapers clearly shows that the public of Vienna could not easily accept if a spectacle that was performed within the Hofburg excluded the general public.

The Hofburg stood out from its urban surrounding by topographical and architectural means which can be interpreted as borders. But these borders had a certain fluctuating character that was visible, for instance, in the way the different facades of the Hofburg were articulated. The Hofburg shared certain infrastructure (and their symbolic meanings) with the public, such as the city's fortifications, and it reached out into the public sphere via satellite buildings, or even allowed public space to expand into courtly territory, as in the case of Josephsplatz. In these zones of overlap court interests and public functions coincided, as on the Inner Hofburg Square (transport and traffic), in the Redouten halls or the winter riding school (cultural events and entertainment), or in the different collections (library and museums), thus dissolving the spatial borders between court and general public. Diverse types of publics were allowed to share imperial spaces mainly because of the functions and uses of these spaces. Thus, the Hofburg was part of the city of Vienna, both topographically and in regard to function.

But as far as the social hierarchy in the Habsburg monarchy was concerned, a strict distinction was always maintained and obviously had to be maintained to identify clearly where the sovereign resided. Where clear borders existed they acted more as social limitations than as visual lines of separation: it was absolutely unthinkable that ordinary subjects who did not belong to the presentable aristocracy, who did not own a high military rank or a post as a minister, could cross this border and find themselves on the invitation list for the famous court balls or balls at court. Thus the Habsburg court could keep its traditional exclusivity, but in most cases only in social terms. The Hofburg therefore was two things at one time: on one hand the central building of the state with prominent institutions that served the interests of the public; this general public was well aware of this character, and developed a certain understanding of it and a will to make use of it. And on the other hand the Hofburg was an elitist space as the home of the ruling monarch who had to maintain his prominent position as the head of the state. It may seem to be a rather difficult task to meet both demands; perhaps it was due to this situation that the borders of the Hofburg showed such an ambivalent character. The Hofburg maintained its exclusive aura as the physical frame of the imperial court, although not primarily by means of architecture. And yet it formed an inseparable part of the city of Vienna; already in the early eighteenth century, the Hofburg was described as the most outstanding ornament of Vienna (even if its architectural shape was regarded as indecent for an imperial palace), and perhaps most telling, in street directories the Hofburg was listed as house number 1.

REFERENCES

Anon. 1880. "Das Caroussel in der Hofburg," *Deutsche Zeitung*, 16 April, Extra-Beilage, 1–8.

Benedik, C. 1993. *Die Redoutensäle der Wiener Hofburg—Kontinuität und Vergänglichkeit*, exhibition catalogue Kulturkreis Looshaus and Albertina Vienna 1993.

Benedik, C. 2008. *Die Albertina. Das Palais und die habsburgischen Prunkräume*, Vienna.

Bericht s.d. *Bericht über die erste allgemeine österreichische Gewerbsprodukten- Ausstellung im Jahre 1835*, Vienna.

Bucher, B. and Weiss, C. 1870. *Wanderungen durch Wien und Umgebung*, Vienna.

Dreger, M. 1914. *Baugeschichte der k. k. Hofburg in Wien bis zum XIX. Jahrhundert* (Österreichische Kunsttopographie 14), Vienna.

Eberle, L. 1911. "Wien als Festung. 1530–1740," *Geschichte der Stadt Wien* 4, Vienna, 218–82.

Feichtinger, J. 2010. "Auf dem Zauberhaufen. Der Burgravelin und die Funktionalisierung des Gedächtnisses an den Entsatz Wiens von den Türken 1683," in *Österreichische Zeitschrift für Kunst und Denkmalpflege* 64 (2010), 108–15.

Freschot, C. 1705. *Relation Von dem Kayserlichen Hofe zu Wien* [...], Cologne.

Holzschuh-Hofer, R. 2010. "Radikal elitär oder schlicht bescheiden? Zur Ikonologie der Wiener Hofburg im 16. Jahrhundert unter Ferdinand I.," in Paravicini, W. and Wettlaufer, J. (eds) *Vorbild—Austausch—Konkurrenz. Höfe und Residenzen in der gegenseitigen Wahrnehmung*. 11. Symposium der Residenzen-Kommission der Akademie der Wissenschaften zu Göttingen (Residenzenforschung 23), Ostfildern, 257–73.

Jeitler, M. 2014. "Der Burgplatz ('In der Burg')," in Karner, H. 2014, 184–7.

Karner, H. 2014. *Die Wiener Hofburg 1521–1705. Baugeschichte, Funktion und Etablierung als Kaiserresidenz* (Veröffentlichungen zur Bau- und Funktionsgeschichte der Wiener Hofburg 2), Vienna.

Küchelbecker, J.B. 1730. *Allerneueste Nachricht vom Römisch-Kayserl. Hofe* [...], Hanover.

Lhotsky, A. 1941. *Die Baugeschichte der Museen und der Neuen Burg* (Festschrift des kunsthistorischen Museums zur Feier des fünfzigjährigen Bestandes, part 1), Vienna-Horn.

Lhotsky, A. 1941–5. *Die Geschichte der Sammlungen* (Festschrift des Kunsthistorischen Museums zur Feier des fünfzigjährigen Bestandes, part 2), Vienna-Horn.

Lichtenberger, E. 1978. *Die Wiener Altstadt*, Vienna.

Lippmann, E. 1881. *Ein practischer Führer für Fremde und Einheimische in Stadt und Umgebung*, Vienna.

Mader-Kratky, "Über das Glacis. Erste Überlegungen zu einer Stadterweiterung Wiens," in Nierhaus, A. (ed.) 2015. *Der Ring. Pionierjahre einer Prachtstraße*, exhibition catalogue Wien Museum Vienna 2015 (forthcoming).

Mayerhofer, J. 1968. "Von der Hofbibliothek zur Nationalbibliothek (1899–1922)," in Stummvoll, J. (ed.) 1968. *Geschichte der österreichischen Nationalbibliothek. Erster Teil: Die Hofbibliothek (1368–1922)*, Vienna, 497–637.

Mrazek-Schwab, E. 1968. "Die Francisco-Josephinische Ära bis zum Ende des 19. Jahrhunderts (1845–1899)," in Stummvoll, J. 1968. 419–96.

Petermann, R.E. 1925. *Historische Spaziergänge in Wien. Die Wiener Burg*, Vienna.

Realis (Gerhard Dützele von Coeckelberghe). 1846. *Die kaiserliche Hofburg in Wien. Ein Wegweiser für Fremde und Einheimische*, Vienna.

Rosenauer, A. (ed.) 2012ff. *Veröffentlichungen zur Bau- und Funktionsgeschichte der Wiener Hofburg 1–5* (five volumes: Telesko 2012 and Karner 2014; other three forthcoming), Vienna.

Schneider, K. 2010. "Hofgesellschaft und Hofstaat," in Rumpler, H. (ed.) 2010. *Soziale Strukturen* (Die Habsburgermonarchie 1848–1918 9, 1/2), Vienna, 1327–48.

Schwarz, W.M. 2010. "Bewegungsspuren. Zur Kritik an der Stadtbefestigung im 18. Jahrhundert," in *Österreichische Zeitschrift für Kunst und Denkmalpflege* 64 (2010), 128–33.

Schweickhardt von Sickingen, F.X. 1832. *Darstellungen der k. k. Haupt- und Residenzstadt Wien*, Vienna.

Telesko, W. (ed.) 2012. *Die Wiener Hofburg 1835–1918. Der Ausbau der Residenz vom Vormärz bis zum Ende des 'Kaiserforums'* (Veröffentlichungen zur Bau- und Funktionsgeschichte der Wiener Hofburg 4), Vienna.

Telesko, W., Kurdiovsky, R. and Sachsenhofer, D. 2013. "The Vienna Hofburg between 1835 and 1918—A Residence in the Conflicting Fields of Art, Politics, and Representation," in *Austrian History Yearbook* 44 (2013), 37–61.

Tietz, C. 1910. "Der Kampf um Alt-Wien," in *Kunstgeschichtliches Jahrbuch der k. k. Zentralkommission* 4 (1910).

Till, R. 1959. "Die Basteien in zeitgenössischen Schilderungen," in *Wiener Geschichtsblätter* 14 (1959), No. 1.

Weigl, A. 2003. "Frühneuzeitliches Bevölkerungswachstum," in Csendes P. and Opll F. (eds) 2003. *Wien. Geschichte einer Stadt*, vol. 2: *Die frühneuzeitliche Residenz (16. bis 18. Jahrhundert)*, Vienna, 109–31.

6

Pier 21 and the Production of Canadian Immigration

David Monteyne

> *I says, 'There's such a terrible oppression coming over me. I'm afraid I can't go up those steps.' So it took a few minutes for me to get up those steps, but I felt, inside, that there was, uh, an oppression—like there's gonna be a lot of suffering there. And there was a lot of suffering. But there were a lot of joys, too ... [T]hey came down from the hand baggage inspection, and then the children were all in the—babies, they had little cribs for the babies and, uh, many a time I had to go and sing to a baby in Italian. [Laughs] Trying to pacify it.*
>
> *Sister Salvatrice Liota, social worker at Pier 21 (Liota, 1988)*

I open with this quotation from Sister Salvatrice not just to get the reader's attention with tales of suffering and joy. Rather, this oral history of part of the immigration process introduces key players and questions that this chapter seeks to explore. To begin with, first appearances can be deceiving. Like Sister Salvatrice, my first assumption on becoming familiar with photos of Pier 21, the Canadian immigration station in Halifax, Nova Scotia, was that this was a place of suffering. I surmised that, by passing through this factory-like building erected in the late 1920s, hapless immigrants were subjected to a Fordist project of the state, by which citizens were produced from un-formed subjects. An assembly line comprising interviews, inspections, and approvals would discharge new Canadians to be delivered by train to distribution points across the country. Some defects would be culled—quarantined or deported.

Certainly, this was a place where people were processed and perhaps transformed too. The Canadian government and other stakeholders hoped it would be seen as a place of national and rational efficiency where the borders of the polity were protected carefully, while the ranks of healthy immigrant laborers (of the right sort) consistently were replenished. To this end, the building combined elements of an industrial warehouse building with those of a typical government building of the time. Long, plain, steel-clad façades of the warehouse spaces for Piers 21 and 22 flank an emphasized central portion that is slightly taller (due to a parapet) and is clad in red brick with masonry details. The overall exterior effect

6.1 Pier 21, Halifax. Streetside façade today.

is symmetry and order, from the flanking wings to the moderately neoclassical elevation of the central portion with its pilasters and cornices. Inside, ordered ranks of benches in the waiting area, and circulation controlled by various barriers and thresholds, reinforce an impression of the efficient sorting of people. But the actual spatial practices of officials, immigrants, and social workers like Sister Salvatrice add a required layer of humanity to the narrative of efficient gate keeping. The very presence of nuns and other volunteers, who worked as intermediaries between officials and immigrants—translating, guiding, steering, serenading, and otherwise comforting newcomers—suggests that passage through Pier 21 was hardly a linear or binary process. In contrast to the abstract line of a national border implicit in the metaphor of a gateway, the architectural border space of Pier 21 was layered and interwoven, even chaotic.

This condition of turbulent flows at the architectural scale is evocative of the broader geographical flows of migrants. An insight of the transnational immigration studies of the past two decades (Gabaccia and Ruiz, 1990; Spickard, 2011; Yans-McLaughlin, 1990) is that there has always existed an ebb-and-flow of migrants between locales, affected by economic and other conditions; and that these flows were only rarely binary ones between one sending nation and another receiving one. So, for example, between Canadian Confederation (1867) and the end of the Second World War, close to half of all immigrants to Canada left for various reasons:

6.2 Sister Salvatrice Liota helping immigrants at Pier 21. Immigrants in Immigrations Hall 3, 1963. Canadian Museum of Immigration at Pier 21 (DI2013.838.2.c).

they returned home voluntarily, were deported, or continued their journeys, usually to the United States (Kelley and Trebilcock, 2010). This flow south across the border, a perpetual problem for Canadian officials during that period, points to the regional character of migration. Broadly, North America was the destination, rather than Canada specifically. Similarly, the origin of European immigrants often was not a nation state, but a portion of an existing or former empire. While national citizenship (or the lack thereof) was sometimes used to erect barriers to admission, Canadian officials typically wrote and implemented policy according to broad and essentialist social, ethnic, or regional categories (e.g., lone females, Galicians, "Mediterraneans"). They also used regional categories to establish hierarchies of preference for immigrant types; the farther removed from Britain and the land, the lower in the hierarchy one was placed. For example, Canada was always wide open to British agriculturalists, who remained at the top of the hierarchy for a good century. Depending on economic and demographic factors in a particular period, Canada would be rendered more or less open to lower levels of its immigrant hierarchy, that is, to southern or eastern Europeans, or to urban artisans.

By the late nineteenth century, immigrants arriving in Halifax or other eastern Canadian ports had been already assessed and selected (often by the transportation company carrying them) as admissible types according to current policy. In the first decades of the twentieth century, pre-screening abroad intensified as admissibility requirements ostensibly became stricter. By 1925, when medical screening was required for almost all immigrants prior to embarkation for Canada, different classes of arrivals might need landing money, passports, and proof of sponsorship or occupation, all things acquired in their home countries (Kelley and Trebilcock, 2010, pp. 188–90). Their processing in port buildings like Pier 21 confirmed their admissible status, their identity, health, and the legality of the contents of their luggage. The general assumption of the immigrants' admissibility allowed for a certain flexibility within the border space bounded by the building envelope—a flexibility difficult for us to understand today, in an era characterized by stricter border protocols. At the same time, architectural space maintained and activated the hierarchies of preference with, for instance, separate facilities and processes for British and "foreign" immigrants, single women migrants, families, and so on. In its operational years from 1928–1971, Pier 21 sorted, shuffled, sifted, and sometimes sequestered migrant bodies that were required to support the national economy and Canadian claims to sovereignty in North America. The production of space at the architectural scale of a building like Pier 21, then, can be understood as one of many moments contributing to the production of transnational space. Since 1999, the building has continued to participate in this production as a museum of immigration.

THEORETICAL CONTEXT

To achieve a comprehensive understanding of the history of Canada's immigration architecture—or of architecture anywhere—it is necessary to study both the formal strategies of governments and institutions and the informal individual and group tactics for shaping space (de Certeau, 1984). In particular, this project on immigration architecture developed out of my reading of cultural theorist Henri Lefebvre's *The Production of Space* (1991). Lefebvre calls for a history of social space encompassing the study of three elements: the spaces conceived and constructed by architects and their clients; the cumulative lived experience of people in these spaces; and, finally, the ongoing, ephemeral spatial practices of people. Of Lefebvre's three elements of social space, it is the architecture of professional designers and powerful patrons that has been the traditional focus of architectural historians. As a corrective, in recent decades vernacular architecture historians have become increasingly sophisticated at researching the lived experience of everyday spaces. Using material culture, popular images, legal documents that establish precedents of use, and other sources, these latter scholars have reconstructed habitual activities and meanings in space (Breisch and Hoagland, 2005; McLeod, 1997; Upton, 2002). Thus, what Lefebvre calls "lived experience" can be understood as a layered storehouse of spatial memory and knowledge drawn upon in everyday life, and which allows individuals or social groups to negotiate and make use of cities, buildings, and other built environments. But it is the ongoing negotiation and use of space that constitutes spatial practice: a mode of bodily action that is framed within existing built environments, and conditioned by previous lived experience, but is not dictated by them. The practice of space can be habitual or conforming to given paths, or it may be improvisational and resistant within material conditions of possibility (Upton, 2002). Spatial practices, often un-choreographed and unrecorded, have eluded historians, few of whom have attempted to research this element of Lefebvre's triad for understanding social space. Anthropologists, geographers, and architectural theorists can study spatial practices in contemporary situations. But ethnographic interviews, cognitive mapping, or observational techniques are not available to the historian of space, who is limited to archival sources and architectural remains. Thus, spatial practice in historical contexts has been relatively difficult to assess.

To make a further distinction between lived experience and spatial practice, I refer back to the opening anecdote of Sister Salvatrice Liota. On her first visit to Pier 21, she approaches the street side entrance as one would approach any building (Liota, 1988). Countless times in her past lived experience, Sister Salvatrice has approached buildings, ascended the front steps, and passed through a door to the interior. "Entering" is a habitual action in relation to buildings. No doubt, there are different conditions of entering: a dozen broad marble steps leading through a Corinthian colonnade and carved oak doors effects a different representation of space than a narrow alley leading to a basement door. The entrance to Pier 21 falls somewhere in between these two extremes: raised a few steps off the ground and differentiated in materials from the bulk of the building—its grey limestone surrounds contrasting with the red brick of the building—it is nonetheless a fairly utilitarian portal. Yet, the Sister's habitual progress is interrupted by her sense or feeling of "a terrible oppression" associated with the building. We can imagine her clutching the railing or her companion for support. Without this pause in her approach, Sister Salvatrice's entry would have progressed along habitual lines through rehearsed bodily movements, a spatial practice that would have been forgotten instantly, passing into the storehouse of lived experience. But instead, because of her affective pause on the threshold, this break from habit, she later recalls this as a meaningful moment associated with that specific architectural space. Due to the Sister's recording of it, the historian has the rare opportunity to distinguish a spatial practice at the juncture of lived experience (everyday life) and architecture (conceived space).

Others who have taken on the challenge of documenting and analyzing historical spatial practices often approach the research through a conception of deviance: however momentary, deviant behaviors might get documented in police records, local bylaws, alternative media, or other literature that may be read "against the grain" by scholars. Iain Borden's work in *Skateboarding, Space and the City* (2001) is perhaps the most consistently Lefebvrian history of spatial practice. Among social historians, George Chauncey's evocative analysis of *Gay New York* (1995) reveals how everyday spaces were used in unofficial and improvisational ways. Other historians of space (Arnold and Sofaer, 2008; Sewell, 2011; Tolbert, 1999) have taken seriously more normative practices sometimes recorded in subjective first-person sources such as memoirs, diaries, letters, and oral histories. Both of these streams of inquiry have provided models for the present inquiry into the spatial practices of Canadian immigrants. The official processes of immigration to Canada cast certain immigrant spatial practices in a deviant mold; the design of immigration architecture, and the intentions for it indicated by federal government documents offer information about the expectation that arrivals may include excludable people or goods. Meanwhile, first-person sources offer the stories of how ordinary immigrants and others negotiated these new Canadian spaces.

Across Canada, many first-person sources from the late nineteenth and twentieth centuries have been preserved in libraries and archives, often now digitized to be searchable. Many of them have been written or spoken by immigrants reflecting on their migration, a period during which they were especially alert to spatial change. As social historians (Elliot, Gerber and Sinke, 2006) have noted, first-person accounts proliferate in relation to unusual circumstances such as immigration, much more so than in recording daily life situations. Because immigration represents such a dramatic change in space, immigrant memoirs often describe and analyze Canadian spaces according to differences perceived by them. The spatial practices of immigrants contributed to the production of these spaces, which become necessarily a hybrid product resulting from the mix of Old and New worlds, customs and improvisations.

This case study of Pier 21 elucidates a method for embedding spatial practices in the represented space of architecture; lived experience of spaces in the past can be seen to guide spatial practices of immigrants in the historical present. It is my contention that the negotiated character of the processes taking place in the building would be missed by symbolic or functionalist analyses that might see an innocuous warehouse building designed for efficient passage of people and cargo; merely a gate between self-contained sending and receiving societies. The study of spatial practices tells us something about architecture, but also about the space of the frontier: both the abstraction of the frontier as a political boundary, and the performed space of it. The flows and counterflows made evident by this detailed delineation of spatial practices map onto transregional movements of migrants, and also evoke the human interactions with space and power not viewable at the intercontinental scale.

PIER 21

We might assume that a federal government architecture used for a specific purpose and forming a symbolic entry point for thousands of arriving immigrants, tourists, and returning Canadians would put forth a consistent and monumental representation of space. For example, Ellis Island's Main Building for the United States' Bureau of Immigration in New York clearly represents the idea of arriving at a significant border crossing. Immigrants for processing would transfer from their ocean liners to a ferry that would deposit them on Ellis Island. Set back from the slip to emphasize the formal, axial approach of the immigrant, the Main Building's front steps rise through arches framed by Classical and national symbols carved in masonry, before ultimately depositing the immigrant in the impressive great hall, a vaulted basilica-like interior festooned with American flags. In contrast, ocean liners would slip up directly alongside the anonymous industrial façade of Pier 21,

6.3 Ship docking at Pier 21. "Welcome home to Canada." Canadian Museum of Immigration at Pier 21 (DI2013.1027.1).

indistinguishable from the adjacent Pier 22 which was used solely for cargo. The ship's gangplank would attach directly to a second floor breezeway that led into the immigration facility. Only a narrow quay divided the ship from the long wall of the pier buildings, which served as a screen separating arrivals from the city and country beyond.

Despite the common characterization of Ellis Island as the "gateway to America," Pier 21 more closely resembled a gateway through this screen, though as I have suggested above, it did not function so simply. As an island, the Ellis facility was a space apart, disconnected from the mainland of the United States; as others have noted, its island situation contributed significantly to its understanding as a heterotopia of deviation (Dolmage, 2011; Hetherington, 1997). Moreover, the spatial syntaxes of the two contrasting situations indicate added transhipments and thresholds in the Ellis Island entry process when compared to Pier 21 (Markus, 1993). Combined with the differing aesthetics of the two facilities, these spatial arrangements communicate qualities and meanings of the spaces. That is, Pier 21 seems to suggest a more literal and precise border between Canada and not-Canada, rather than a series of "border crossings," each one moving the immigrant slightly closer to the Promised Land. This unassuming and seemingly direct approach to the architectural boundary of Canadian space was typical of the similar transshipment points found in other eastern Canadian ports receiving European immigrants, specifically Quebec City and St John's, New Brunswick. At those ports, as in Halifax, ships docked and trains stopped immediately adjacent to the pier buildings, so the buildings themselves became the border space. However,

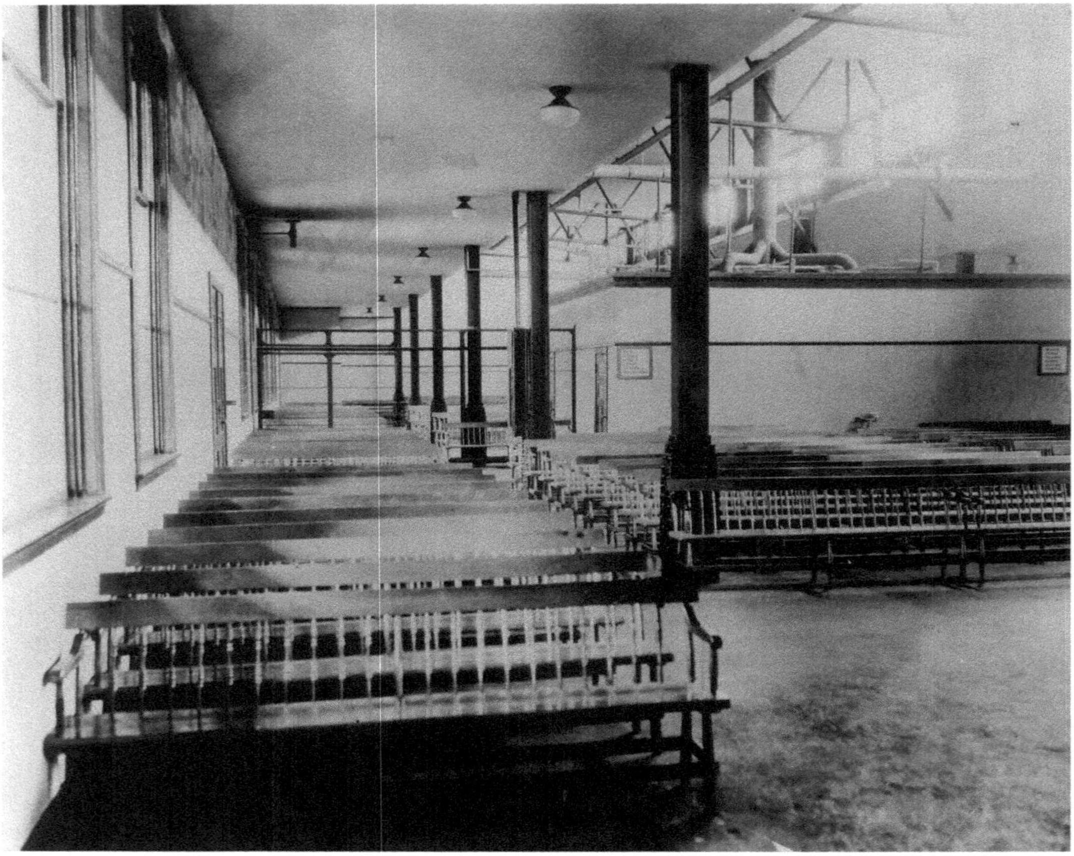

6.4 Pier 21, interior view of waiting room. Reproduced by permission of Halifax Port Authority, Canada.

despite its seeming fence-like quality, inside Pier 21 there was no particular point or line where an immigrant stepped across the national border. Rather, as I will discuss below, the border space was stretched and manipulated by the range of actors within the building.

If the building seems designed to express the efficiency of its architecture, Pier 21 actually became an immigration station somewhat inadvertently. The building was never envisioned as such, and the social space of the immigration processes was inserted into a warehouse space originally designed and built by railroad engineers as part of a massive project known as Ocean Terminals, a deep sea port constructed on fill beginning in World War I. In fact, the Department of Immigration and Colonization resisted being put in Pier 21, arguing for the continued occupation of their previous quarters in Pier 2, purpose-built for them just before the War (even though its surroundings were devastated in the great Halifax explosion of 1917, which occurred at Pier 6). Local immigration officials had no desire to move. But the port authorities decreed that European arrivals would now dock at the new Pier 21, which was located adjacent to the Union Station and railway hotel being built at the same time. Tracks were laid to service the new terminal, and the so-called "colonist trains" would henceforth pick up the immigrants directly from Pier 21, on a spur line.

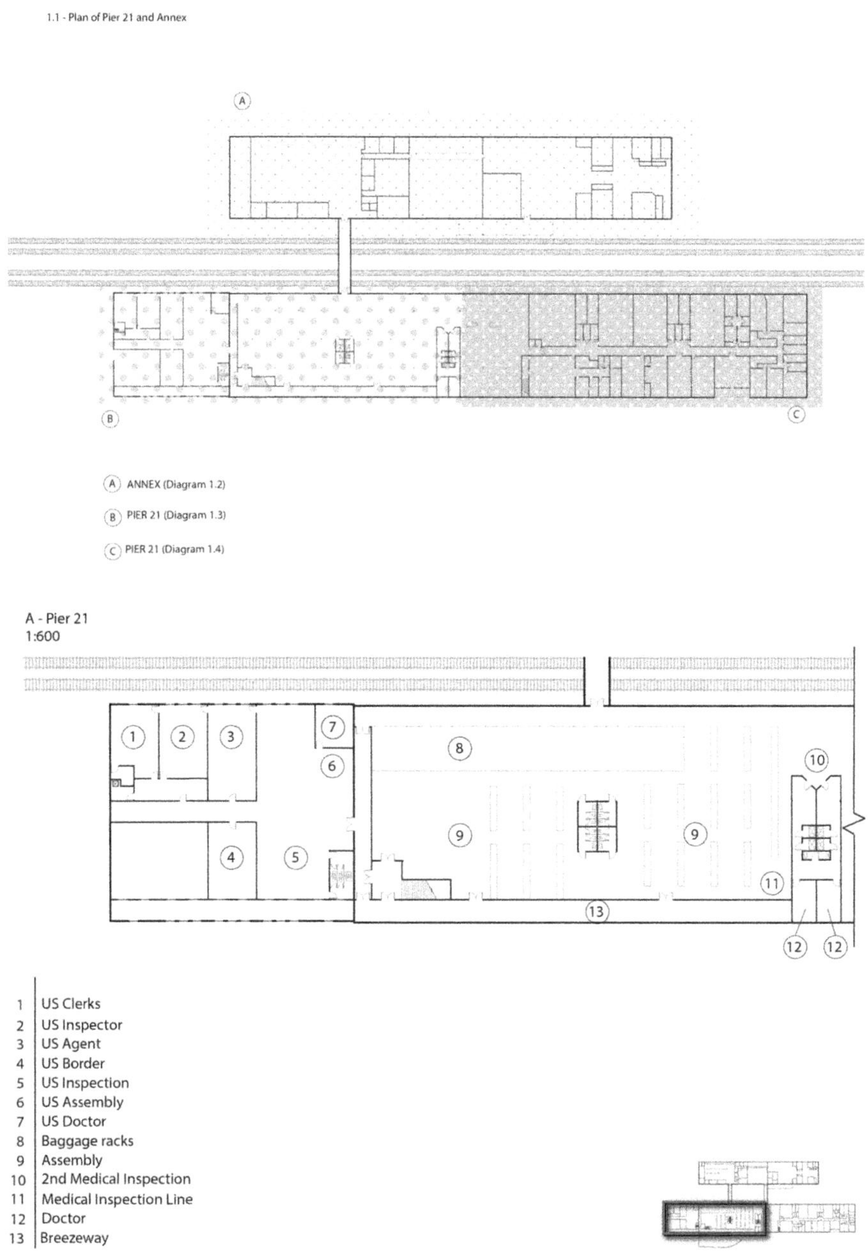

6.5 Four floor plans of Pier 21 and Annex. Drawn by Kendra Kusick (*continued overleaf*).

B - Pier 21
1:600

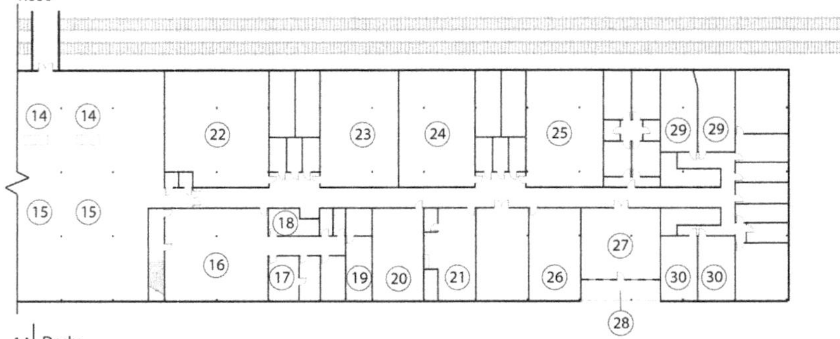

14 Desks
15 Civil Inspection Line
16 General Office
17 Agent
18 Stores
19 Boardroom
20 Detention Sailors
21 Kitchen
22 Detention British Men
23 Detention Foreign Men
24 Detention Foreign Women
25 Detention British Women
26 Cubicles
27 Recreation Room
28 Balcony
29 Ward Foreign Men
30 Ward Foreign Women

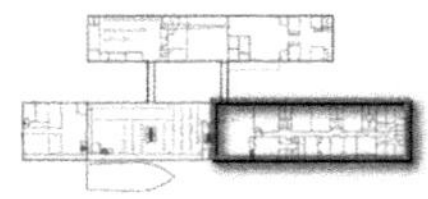

C - Pier 21
1:600

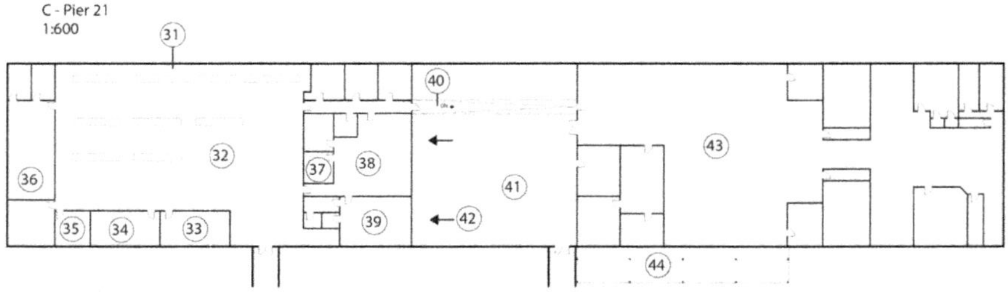

31 Social services
32 Waiting room
33 CPR Ticket office
34 CNR Colonisation
35 Money changer
36 CNR Ticket office
37 Kitchen
38 Red Cross
39 Nursery
40 Ramp
41 Baggage room
42 Baggage room continued beneath
43 Waiting Room
44 Train platform

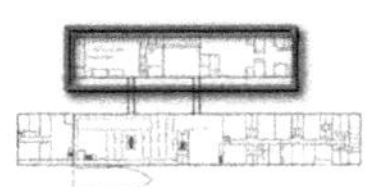

6.5 *continued*

Construction was well under way when the federal Department of Immigration and Colonization was forced to plan for occupying Pier 21 in 1928. The building was designed to be a cargo shed, as one section of a series of sheds along the waterfront; the Department was offered a large, unfinished, open space on the second floor. Using partition walls under the open, steel-truss roof structure of the shed, officials carved up the space into offices, waiting rooms, an infirmary, and detention center. It was not enough space to accommodate the full immigration process, so they insisted on the construction of an annex across the railroad tracks. Built at the same time, the Annex housed customs and baggage handling, railway ticketing, provisioning, and the social service agencies, including a nursery operated by the Red Cross. Pier 21, therefore, was an industrial transshipment point, a portion of which was adapted to the processing of human cargo—a sort of Customs clearance for people. But this "cargo" actively participated in and reflected on the experience of arriving at and passing through the building.

Ships pulled up to and docked next to the unadorned façades of Piers 21 and 22, a continuous wall more than 1,000 feet in length. If immigrants to Pier 21 seemed to be arriving at Canada's back door, feelings of dread were tempered by ones of relief and excitement, as related by Dutch immigrant Akky Mansikka:

> *I remember coming in and seeing these islands over here and they were tree-covered and the first time in my life seeing brilliant—all the greens of the evergreens and the brilliant yellow, brilliant red—I mean I thought it was magic land ...*
>
> *I remember the building ... it was dark and dingy and it was like we were going through warehouses. I mean, I had no idea it was warehouses I guess. It seemed like millions of people in lineups and barricades to keep the lines straight, [people were] shuffling along like, I mean, to a little kid it was like days and days in here. (Mansikka, 2000)*

During the period of Pier 21's operation, immigrants were pre-screened abroad, so processing at the border was mainly a formality. But Department of Immigration and Colonization officials made identity checks and clean-bill-of-health confirmations as formal as possible. Benches to seat almost 1,000 immigrants were arrayed, church-like, in two waiting areas, facing the desks of officials. Closed doors led to unknown rooms, into which select immigrants were taken for further examination. The high ceiling of the warehouse space, receding above the partition walls, suggested untold areas where immigrants might be detoured by bureaucracy. Officials wore uniforms and caps, while security guards and police stood by, and chain-link baggage cages lent the air of a prison, as noted by immigrants and officials alike in their oral histories. In particular, this atmosphere of governmental power was seen to be problematic when welcoming refugees from World War II or from behind the Iron Curtain. Another immigrant narrative offered an alternative reading, emphasizing the industrial character of Pier 21. Max de Bruyn, who immigrated from Holland as a child, recalled the building as a "warehouse," with immigrants assembled in "a processing area that resembled

6.6 Interior view of Pier 21, with waiting room and baggage cages. Reproduced by permission of Halifax Port Authority, Canada

a stockyard. Herded and harangued like cattle at a Friday auction" (Charon, 1983, p. 126). The organization of space and the spatial practices of officials at Pier 21 were intended to portray a controlled passage toward an end point.

As formal and channeled as this sounds, when a shipload of immigrants animated the space it was total chaos: multiple languages being spoken at once, babies crying, restless children, and worried adults. Moreover, there was no hermetically sealed border line that can be demarcated within the building. Though the social service workers from various churches and other organizations were based across the bridge in the Annex, they had access to most areas. They often met immigrants at the foot of the gangplank, identifying and offering assistance to those of their denomination or ethnicity, even before the new arrivals entered the waiting rooms or met with anyone in an official capacity. Babies, children, and even mothers often were whisked off to the Red Cross nursery in the Annex, seemingly across the border, while other members of their families awaited processing by Canadian officials.

When the wait was over, immigrants received perfunctory medical exams from the resident physician to ensure that they had not contracted something since their health inspection by doctors (under contract to the Canadian government or transportation companies) at their point of origin. Following this, immigrants would meet with Immigration officials at the desks to go over their papers. When necessary, they were joined by translators—often the Sisters or other social service volunteers, or sometimes a Haligonian tracked down in the city and brought to Pier 21 to help out with some of the rarer languages. Sister Florence Kelly (an Irish nun who spoke fluent German and Italian) admitted in her oral history to having steered vulnerable immigrants away from "nasty" Immigration officers. At the foot of the gangplank one day she also recognized an illegal immigrant who had been rejected a year earlier—under a new identity he successfully entered Canada on his second try (Kelly, 1998). In an interview many years later, one illegal immigrant told of arriving at Pier 21 as a stowaway, and shimmying down the ship's hawser in the

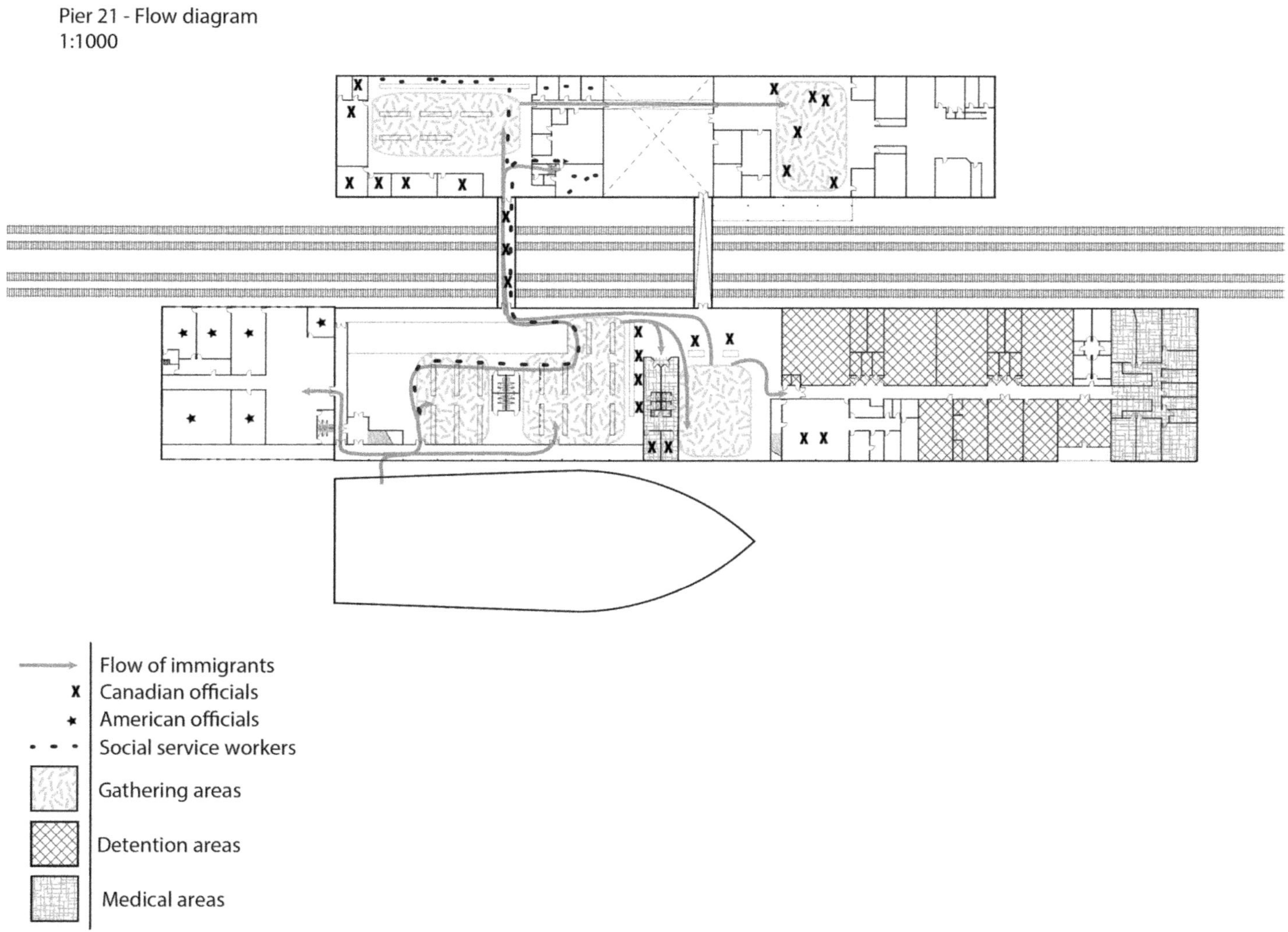

6.7 Pier 21, spatial arrangement and flow diagram. Drawn by Kendra Kusick.

6.8 Immigrant interviewed by immigration official and translator. Unidentified Customs Officer with family, 1963, Canadian Museum of Immigration at Pier 21 (DI2013.1362.4).

night and off into the harbour to hop a train (Charon, 1989, p. 59). Stories like these suggest that the formal border represented by Pier 21 was a series of negotiated spaces, highly contingent upon the range of practices producing them.

Some small percentage of immigrants were detained for health or bureaucratic reasons in facilities just off the main waiting rooms. A few were deported. The infirmary and detention center have their own interesting history. Two nurses lived full-time inside Pier 21, coming and going as they pleased between Canada and their apartment, passing daily through the liminal space of detention. Patients used the telephone in the nurses' apartment, which also was carved from the overall volume of the warehouse space by use of partition walls. Meanwhile, detainees slept in gender- and ethnicity-separated dormitories, and deportees were kept in cells—but they all dined together and had access to the recreation room and balcony overlooking the quay. Detainees (though not deportees) were allowed to exit Pier 21 to visit the city of Halifax; some found temporary work in the city. This was a markedly different experience from the isolation of Ellis Island. Some immigrants lived in Pier 21 for months while their paperwork was completed or they awaited Department of Immigration decisions. These stories of nurses and detainees add a further layer of spatial practices, with domesticity woven into the industrial and governmental architecture of Pier 21.

In certain periods, the flows into the country through Pier 21 were reversed. For instance, during the 1930s Depression, when European immigration all but stopped, Pier 21 became a holding center for political and economic deportees awaiting transportation. Most notoriously, Pier 21 received national attention in 1932 during the detainment there of the "Halifax 10," men accused of communism and sedition. All 10 were immigrants to Canada, though many had been in the country for 20 years at that point. They had been arrested in May Day raids across Western Canada, and spirited out of their home communities by night. The men were sent to Halifax to separate them from their local support networks in the West, while a Department of Immigration board of inquiry, and then ultimately the Supreme Court of Canada, determined their deportation cases. In these cases, the detention area of Pier 21 became a steel and concrete prison under 24-hour guard, with no exercise yard or town privileges. Photographs of the 10 men published at the time by *Labor Defender*, and "taken under greatest difficulties" through the second floor windows of the detention area, cast an eerie aspect on the space of Pier 21. The bars on the windows cast shadows across the faces of the men, who seem to be posing for the surreptitious photographer in the detention area's lavatory. The "Halifax 10," and many others during the 1930s, were deported to Europe for reasons of politics or indigence, sometimes to face horrible fates in fascist countries (Kelley and Trebilcock, 2010, pp. 246–8; Roberts, 1988, pp. 135–41).

Returning to the narrative of entering the country though Pier 21, immigrants who were not removed from the general flow into the eddies of detention, were passed from the Immigration desks to a corridor that connected them with the baggage cages, and then the enclosed bridge to the Annex. On the bridge they passed through their first Customs inspection, that of their hand luggage. From here, they crossed into the Annex where they might retrieve children who had preceded them into the country, unexamined. If they had not done so already, the immigrants could change their money or purchase train tickets for the next stage of their journey to whatever remote town or plot of land they were headed to. At their counters, social service workers helped immigrants post letters home and write grocery lists for the small shop located in the Annex. The "colonist trains" did not have dining cars (they barely had heat or beds), so immigrants required supplies for the long ride across the Canadian shield and prairie. Despite being now in Canada, most of them had a long way to go. Still, they would remain under the paternal care of the state as they continued on into the interior of Canada. For example, other federal government buildings dotted their route, offering free accommodations and advice in transportation hubs like Winnipeg and Edmonton, and in small towns at the end of the railroad tracks near available land.

A long ramp in the Annex led down to track level where there was a waiting room for the trains, and the baggage room. While they were being processed upstairs, the immigrants' baggage had been transferred from the hold of the ship through the ground floor of Pier 21 and into the Annex baggage room. Here the immigrants claimed their trunks and other large items, but not before they were inspected by Customs officials, and sometimes by representatives from the Department of Agriculture such as R. David Gray, who provides a brief review of the whole process, in an interview with Steve Schwinghamer:

> *DG: And then the other part of course is where the, where the immigrants were taken upstairs, interrogation rooms and sitting on benches.*
>
> *SS: What were those rooms like?*
>
> *DG: Those were just more or less kind of sterile areas. There were just wooden benches. They weren't very colourful. The walls were, you know, just plain walls. There was no real colour involved. It wasn't the kind of place that I would have expected that Canada would have to greet the immigrants. And it was a new country they were coming to and so on. It was very, very sterile. And they'd sit there for hours on these long wooden benches waiting to be interviewed and then, they'd come out and go down the catwalk and bring their hand baggage. Of course that's when you'd see them with the hand bags. But down below, the heavy trunks and other stuff would be laid out on the floor by name, by alphabetical—I can remember by A, B, right down and come back the other way. And there would be customs officers stationed all the way along there and they had these large, big suitcases and trunks sitting there with the olive oil cans. Some of them had holes in them and the olive oil would be running out onto the floor. And it would be sticky. The floor was sticky anyway. Sticky from all the previous oils. It must have had a layer of olive oil and grime on it. It was just dirty. Grimy and dirty. The whole building was, you know, like a dungeon in there and you could hear the chatter. I remember the chatter of the wives and the mothers of the Europeans especially from the Mediterranean area.*
>
> *SS: Yeah.*
>
> *DG: They'd be chattering away and giving directions to the customs officers telling them what to do, as opposed to—*
>
> *SS: As opposed to vice versa. (Gray, 2000)*

Homemade wine and spirits were commonly confiscated as well. Immigrants anticipated these inspections and went to great lengths to conceal banned items, such as sinking them in the large cans of olive oil, or even slipping them temporarily into their overcoat pockets! The baggage room has been described by officials and immigrants as a place of chaos, with children running and climbing on piles of baggage, the unpacking and re-packing of trunks and crates, and dragging them out to the platform, and Customs officers trying to keep track of it all. Officers used a system of chalk marks to indicate to security guards that baggage had been inspected. Several immigrants recall finding broken pieces of chalk on the floor and marking their own Xs on their trunks and crates before anyone could inspect them. This was enough to get their baggage, and everything it contained, out the door and onto a train.

The tireless Sisters and others often boarded the trains to distribute free magazines and foreign language newspapers, children's books, ditty bags, and even promotional mini-boxes of Corn Flakes, donated by Kellogg's in a scheme to win over newcomers to the consumer habits of Canadians. Trapped in conversations and reassurances with immigrants when the colonist cars began to leave the station, the Sisters would leap off the moving train before it entered the harbor tunnel and departed Halifax.

CODA

Most immigrants spent only a few hours in Pier 21, between disembarking and departing by train for the rest of Canada. During that time they crossed multiple boundaries that made up the abstract national border. Immigration and customs regulations; language differences; money and transportation dilemmas; the introduction to white bread and other North American delicacies: all were barriers crossed within the architectural space of Pier 21. Others encountered health or political barriers that required them to stay longer in the building. At an abstract, national scale, it is tempting to see Pier 21 and Canada's other immigration sheds as sluice gates channeling the flow of immigrants across a relatively porous border. At an architectural scale, however, the design and practice of space muddies the waters of this metaphor. In the flow of immigrants through Pier 21 there were obstacles creating turbulence and eddies, as well as leaks and evaporations, condensations and rapids, and of course, counter-flows.

How immigrants navigated these waters depended to some extent on their past lived experience of similar and other buildings and interiors which might have given them clues for how to practice newly encountered spaces. As the immigration historian Sucheng Chan (1990) has noted, in "defending themselves" from the inherent risks of encountering new environments and social relations

> *immigrants have often drawn on skills developed in their 'Old World' societies, which were more hierarchical and more rigidly bound by age-old status prescriptions. Although the nature of the restrictions in the 'New World' differed, the existence of limitations per se was something with which immigrants were quite familiar. (1990, p. 61)*

An architecture's representation of space, we might say its message, is registered and received, then interpreted in relation to previous experiences of architectural space. Spatial practice follows, feeds into, and builds upon, this process.

As a government building, and presumably the nation's front door, Pier 21 offered mixed messages and varied interpretations. To the arriving immigrant, the continuous wall adjoining the dock must have seemed an indomitable boundary. Inside, the trappings of authority might have intimidated new arrivals as they sat in pews looking at men in uniform and a giant Union Jack hanging from the ceiling. At the same time, the exposed-truss ceiling aloft over the tops of partition walls must have seemed impermanent and non-monumental. While some waiting rooms would have been everyday public spaces that immigrants were accustomed to, the waiting rooms of government agencies are arranged in ways that emphasize state power over the subject (Adams, 2008; Edelman, 1964). Adults know how to behave in such spaces. But children, less layered and molded by these types of lived experience, and with weeks of pent-up, shipboard energy to expend, brought different, perhaps unruly or disruptive practices into the building. In addition, the nursery and other social services, and food, were on the other side of the legal border, and premature access to the Annex seems to undermine the representation of national space. Finally, the chaos of the baggage room may have allowed immigrants to draw on lived experience of other spaces—for example, markets or workplaces—to successfully negotiate the movement of their families and objects using specific spatial practices, licit or illicit.

For Immigration and Customs officials such as R. David Gray, the space of Pier 21 would have been different again, as their oral histories suggest. Their lived experience may have come closer to the intended representations of space, but also was produced through negotiated practices. If they interrogated immigrants, or confiscated their goods, on behalf of the state, they also sympathized with the new arrivals' feelings of fear and wonder. A number of them commented in their oral histories on the lack of welcome evident in the sterility and industrial character of the building. Since immigrant stories of smuggling banned goods through Pier 21 are so prevalent, one also wonders whether the Immigration and Customs officials were completely ignorant of these spatial practices, or if they looked the other way.

Meanwhile, the participation of volunteers like Marguerite Peters at Pier 21 engaged them in new spatial practices, and exposed them to new representations of space, both local and global. Peters, from an "old Halifax family," stepped out of her gender, physical, and ethnic comfort zones when she began volunteering at Pier 21 with the Catholic Women's League. As she relates in her oral history, the waterfront was not a comfortable place for respectable women, but the immigrants' spatial practices changed the rough and masculine nature of the space:

> *Bleak. Yes, it was, uh, it was certainly new to me, I hadn't been down around the piers or anything like that until this occasion when I started to do the volunteer work down there. And it was—what's to say? It was a pier ... It was, you know, cold, there was no ambiance around ... When you went you made sure you were dressed, you always left your boots on. And sometimes your coat too, as well. ... I'll tell you, when the people came off ... they just seemed to warm the whole place up, you know, and thank heavens they did, but they were so appreciative of where they were and where they were going ... And the kids were so, so excited, you know, and that–they were really something. And, you know, it would be hectic at times, because–and of course, then you had the language barrier. (Peters, 2002)*

Once, when a ship full of immigrants arrived late at night and needed to be processed, her husband insisted on escorting her to the waterfront. When he witnessed Pier 21 come to life, with its money exchange and ticket counters, nursery, canteen, and store, Peters' husband

> *[...] was just astounded. He said it was—to him it was just like a city opening up in the middle of the night. That was the feeling he got. And of course, he hadn't seen the people coming in with a variety—at that time, there were a lot*

> *of Italians then. And they would come in, the children with their dolls, you know, beautiful, beautiful dolls, and the men with the wicker baskets with the wine bottles in them, you know. And their different dress, the whole thing was a real eye-opener, a real education for anybody ... I've never forgotten, that experience that I had down at the port, it was certainly something—one of the highlights in my life. (Peters, 2002)*

This urban sense of the spatial mixing of people and cultures became lived experience which evidently stayed with this couple, influencing subsequent spatial practices.

Let us return at the end, though, to the intermediary, Sister Salvatrice, who read the building's representation as a space of authority and even oppression when she first set foot on the front steps (or is it the back steps of the building that face the city and the rest of Canada?). What lived experience had prepared her for the role she played at Pier 21: interceding with God on behalf of penitents? At any rate, her spatial practices underscored the contingencies of the border space. She passed across it with impunity, a citizen of no nation, subject to a higher power. But really, she did not need divine authority to produce space in this way. Children and mothers, shipping agents, Red Cross workers, amateur translators, nurses, and others interacted across the architectural space of Pier 21, producing a layered and multi-directional border geography.

David Monteyne wishes to thanks Steve Schwinghamer and other research staff at the Canadian Museum of Immigration at Pier 21.

REFERENCES

Oral Histories (unpublished)

Gray, R.D. 2000. Interview by Schwinghamer, S., Pier 21 Oral History Collection. Halifax, NS. CD Recording.

Kelly, Sister F. 1998. Interview by Morrison, J.H., Pier 21 Oral History Collection. Halifax, NS. CD Recording.

Liota, Sister Salvatrice, 1998, interview by Morrison, J.H. Pier 21 Oral History Collection. Halifax, NS. CD Recording.

Mansikka, A. 2000. Interview by Coleman, A., Pier 21 Oral History Collection. Halifax, NS. CD Recording.

Peters, M. 2002. Interview by Tinkham, J., Pier 21 Oral History Collection. Halifax, NS. CD Recording.

Print Sources

Adams, A. 2008. *Medicine by Design: The Architect and the Modern Hospital, 1893–1943*. Minneapolis: University of Minnesota Press.

Arnold, D. and Sofaer, J. (eds) 2008. *Biographies and Space: Placing the Subject in Art and Architecture*. New York: Routledge.

Borden, I. 2001. *Skateboarding, Space and the City: Architecture and the Body*. Oxford: Berg.

Breisch, K.A. and Hoagland, A.K. (eds) 2005. *Building Environments: Perspectives in Vernacular Architecture, X*. Knoxville: The University of Tennessee Press.

Chan, S. 1990. "European and Asian Immigration into the United States in Comparative Perspective, 1820s to 1920s." In V. Yans-McLaughlin (ed.) 1990. *Immigration Reconsidered: History, Sociology, and Politics*. New York: Oxford University Press, ch. 2.

Charon, M. 1983. *Between Two Worlds: The Canadian Immigration Experience*. Dunvegan, Ontario: Quadrant Editions.

Charon, M. 1989. *Worlds Apart: New Immigrant Voices*. Dunvegan, Ontario: Cormorant Books.

Chauncey, G. 1995. *Gay New York: Gender, Urban Culture, and the Making of the Gay Male World, 1890–1940*. New York: Basic.

de Certeau, M. 1984. *The Practice of Everyday Life*. Berkeley: University of California Press.

Dolmage, J. 2011. Disabled upon arrival: The rhetorical construction of race and disability at Ellis Island. *Cultural Critique*, 77 (Winter), pp. 24–69.

Edelman, M.J. 1964. *The Symbolic Uses of Politics*. Urbana: University of Illinois Press.

Elliott, B., Gerber, D.A., and Sinke, S.M. (eds) 2006. *Letters Across Borders: The Epistolary Practices of International Migrants*. New York: Palgrave Macmillan.

Gabaccia, D.R. and Ruiz, V.L. (eds) 2006. *American Dreaming Global Realities: Rethinking U.S. Immigration History*. Urbana: University of Illinois Press.

Harris, S. and Berke, D. (eds) 1997. *Architecture of the Everyday*. New York: Princeton Architectural Press.

Hetherington, K. 1997. *The Badlands of Modernity Heterotopia and Social Ordering*. London: Routledge.

Kelley, N. and Trebilcock, M. (eds) 2010. *The Making of the Mosaic: A History of Canadian Immigration Policy*. Toronto: University of Toronto Press.

Lefebvre, H. 1991. *The Production of Space*. Oxford: Blackwell.

Markus, T.A. 1993. *Buildings and Power: Freedom and Control in the Origin of Modern Building Types*. New York: Routledge.

McLeod, M. 1997. "Henri Lefebvre's Critique of Everyday Life: An Introduction." In Harris, S. and Berke, D. (eds) 1997. *Architecture of the Everyday*. New York: Princeton Architectural Press, pp. 9–29.

Roberts, B.A. 1988. *Whence They Came: Deportation from Canada, 1900–1935*. Ottawa: University of Ottawa Press.

Sewell, J.E. 2011. *Women and the Everyday City: Public Space in San Francisco, 1890–1915*. Minneapolis: University of Minnesota Press.

Spickard, P. (ed.) 2011. *Race and Immigration in the United States: New Histories*. New York: Routledge.

Tolbert, L.C. 1999. *Constructing Townscapes: Space and Society in Antebellum Tennessee*. Chapel Hill: The University of North Carolina Press.

Upton, D. 2002. "Architecture in everyday life." *New Literary History*, 33, pp. 707–23.

Yans-McLaughlin, V. (ed.) 1990. *Immigration Reconsidered: History, Sociology, and Politics*. New York: Oxford University Press.

7

Bordering on Peace: Spatial Narratives of Border Crossings between Israel, Jordan and Egypt

Eric Aronoff and Yael Aronoff

Crossing the border from Jordan into Israel at the Itzhak Rabin/Wadi Araba Border Terminals, travelers encounter a curious sight. Since they are in the Arava Valley, just outside the Jordanian city of Aqaba and the Israeli city of Eilat, where annual precipitation is 3 cm on average, nearly all border functions are performed with travelers standing outdoors (Barzilay, 2000). After proceeding past a series of windows along the outside of a long, narrow one-story building that serves as Jordanian passport control, travelers exit Jordanian territory passing under a portrait of the late King Hussein and his son, the current King Abdullah, while signs wish the traveler "Goodbye" in English and Arabic. Walking across the no-man's land between the terminals, travelers are greeted with a large billboard, visible across the zone, proclaiming "שלום Peace سلام" ("Shalom Peace Salaam") before entering the terminal under a welcome sign in English, Hebrew and Arabic. Upon entering Israeli territory, the traveler is ushered into a large, well-lit room with several metal detectors, for screening both travelers and their luggage, much like the security scans at any airport. (In fact, the border terminal itself was designed and is run by the Israel Airports Authority, and the entire space has the modern feel of an airport terminal.) What is striking about the room, however, is its decoration. Lining three of the four walls are a series of framed photographs: on the left, a series of photographs of the late Prime Minister Itzhak Rabin, at various stages of his career from his youth in the Hagana through his Prime Ministership; on the right, a series of pictures of Rabin with his family. As the traveler moves through the middle of the room, passing through the security screens, the photos on the far third wall come into view: these photos all show Rabin with King Hussein—Rabin and Hussein sharing a cigar; Rabin and Hussein gazing together over the Gulf of Aqaba; several pictures of Rabin, Hussein and US President Bill Clinton together signing the 1994 peace treaty between Israel and Jordan that opened the Arava crossing.

The contrast is striking: between the largely open-air design of the terminals on both sides of the border and this decorated room; between the iconography of the Jordanian terminal focused exclusively on the royal family, and how this space highlights Rabin's relationship with King Hussein and the peace negotiations. What, one is led to ask, is the story being told in here, and why this particular story? How is this border space designed to tell that story to the traveler passing through? To whom is this story being addressed? Is it aimed at every traveler crossing this border, or are certain audiences particularly addressed, and why? And is the story being told here a story reproduced at other border crossings between these two countries, to form some kind of official narrative—or are there variations, planned or ad hoc?

These questions about border narratives are the focus of this essay. Examining the border crossings between Israel and the two neighboring states with which it has open borders, Jordan and Egypt, we analyze the narratives created in these spaces through the arrangement of space, iconography, and signage, as well as the legal elements that also regulate the flow of persons across the borders. These sites, in effect, constitute the first encounter of travelers with the new state about to be entered; as such, these spatial, visual and legal elements combine to create a "story" being told to that traveler (even if that traveler is a member of that community who is returning). That story may be intentional, the result of a conscious effort or policy on the part of the state, or unintentional as the ad hoc reflection of attitudes and ideas expressing themselves through the choices made "on the ground" by border personnel. That story is both about who "they"—the imagined community whose territory the traveler is about to enter—are and what they represent; it also simultaneously is about who "you," the traveler, might be—why you might be there, the relationship imagined between "they" and "you." Like a text, these spaces construct both their ideal "author" and their ideal "reader."

In this way, like many of the chapters in this volume, our approach extends but also differs from much of the scholarship that makes up the recent resurgence of border studies. As many scholars have pointed out, rising attention across multiple disciplines to issues of globalism and transnationalism, as well as cultural studies approaches to concepts heretofore in the domain of social science, have resulted in increased interest in borders (Newman, 2011). Until relatively recently, borders have been approached within the fields of international relations or geography as static, empirical entities, largely in the context of examining relations between states (Sack, 1986; Taylor, 1994; Shapiro and Alker, 1996). More recent theories emanating from anthropology and cultural studies have emphasized the social construction of boundaries as processes for defining personal, group and national identities, through processes of inclusion and exclusion, defining the "self" and the "Other." These approaches have broadened the concept of "borders" to include not only the actual borderline between states, but many other kinds of borders. In this conception, borders are everywhere, and the "border narratives" that constitute them are made up of multiple discourses and texts: newspapers, political speeches, posters, poems, plays, novels, everyday speech that give meaning to the border as the "construction of institutionalized forms of 'we' and the 'other' which are produced and perpetually reproduced in education texts, narratives and discourses" (Newman and Paasi, 1998, p. 196).

But even these examinations of border narratives, like the older tradition of border studies, focus on the construction of mutually constitutive in-groups and out-groups demarked by the border that is now "narrated" into being. They are concerned with the meaning of borders as they are erected: who is allowed inside, who is kept outside, who is allowed to pass through. This approach still (metaphorically) treats borders as a "line," not a "space." Our narrative approach looks at the *way* those who are allowed to cross the borders are *ushered through* that liminal space—a space where, in the case of actual border terminals, narratives of *relation* are constructed *in the experience of moving through them*. These narratives go beyond establishing who is separated and who is connected, to narrating *how and why* they are connected. The organization of the border itself as a spatial experience shapes the meaning of the border crossing. In other words, while there might be a narrative constructing the identities of those within the borders, those without, and those who can cross the border, there is *also* a narrative constructed by the crossing itself—a narrative constructed, programmatically or implicitly, by the arrangement of space and iconography of the border terminals themselves. Thus, we argue that these border crossings construct narratives of relation as sites in themselves, narratives that are experienced in the process of moving through that space.

Our methodology is interdisciplinary, drawing on ethnography, history, political science and cultural studies approaches to borders. Our data was derived from several trips to each border crossing, taking place in the summers of 2010 and 2013, in which we crossed each border, read (and where possible photographed) the spatial design and iconography of the sites, and spoke to personnel at the crossings about their own understandings of the histories and meanings of the sites. We examined each of the four border crossings between Israel and Jordan, and Israel and Egypt. Conversations took place in English and in Hebrew; thus interviews on the Jordanian and Egyptian sides of each crossing were limited by the English language skills of the personnel there. Further data was gathered through correspondence with personnel in the Israel Airports Authority, the administrative body responsible for the design and maintenance of the Israeli border terminals.

We argue that the narratives that emerge at the Israeli border terminals differ significantly from those of the Jordanian terminals, and that in fact the narratives at different Israeli border crossings into Jordan differ from one another as well. On the whole, the iconography and the arrangement of space of Israeli border terminals construct narratives that celebrate and commemorate peacemaking—specific peace treaties, and more broadly the idea of peace between neighbors. This is accomplished through signage, photographs and other iconography conveying these processes. These narratives, moreover, are primarily aimed at travelers entering Israel, as this symbolism is particularly prominent in the entry terminals, and less so at the exiting terminals. Moreover, the precise iconography differs *between* Israeli terminals, as each site seems to address its message to the particular populations that are imagined to be crossing at that site. On the other hand, Jordanian terminals showed less variation in their iconography and arrangement of space, each highlighting the role of the Hashemite monarchy and creating a narrative of national stability and continuity. Finally, the iconography and arrangement of space at the Taba crossing between Egypt and Israel are likewise different. The Israeli terminal also commemorates the establishment of peace between Egypt and Israel, but this commemoration is aimed at Israelis and tourists leaving Israel, rather than Egyptians and tourists entering the country. The Egyptian terminal, like the Jordanian, does not mark the peace process with Israel in any way. Instead, through a kind of monumental architecture and photographs of ancient Egyptian art and artifacts, it highlights the grandeur of Egypt's cultural heritage.

DOORS WIDE SHUT: THE HISTORY OF BORDER CROSSINGS

In order to understand the border terminals between Israel, Jordan and Egypt and the iconography that shapes their meanings, a brief account of their tangled history is required. For in fact Israel *had* no open border crossings with its four neighboring states—Egypt, Jordan, Syria and Lebanon—for the first 31 years its existence as a state. In 1948, the United Nations approved the plan for the partition of what had been the British Mandate of Palestine. Jewish groups accepted the plan and declared the state of Israel; the Palestinian leadership and the four neighboring Arab states, along with Iraq, rejected the plan and declared war. This formal state of war continued through 1979, even between the "hot" wars(the War of Independence of 1948–49; the Sinai War of 1956; the Six-Day War in 1967, and the Yom Kippur War of 1973) with none of its neighbors recognizing Israel's existence. During this period, the potential lines of any possible border shifted dramatically. In 1967 Israel launched a preemptive war against Egypt and Syria, based on intelligence of an imminent attack. As war began with Egypt and Syria, Israel warned Jordan—who after the 1948 war had gained control of much of what would have been a Palestinian state including the Old City of Jerusalem and the West Bank of the Jordan river—to stay out of the war; when Jordanian forces attacked, Israel retaliated with an offensive that pushed to the Jordan River. By the end of that conflict, the Israeli military had taken over the Gaza strip (which had been in Egyptian hands since 1948) and the Sinai from Egypt, East Jerusalem and the West Bank from Jordan, and the Golan Heights from Syria. Israel's offer of "land for peace" deals to all three states were rejected by the entire Arab League, leaving all three territories in Israeli hands—a situation that had profound and tragic effects on all in the region, as well as, for our purposes here, on the shape and location of any potential border crossings between the countries.[1]

---

1 Of course, with a history and context as fraught as the Middle East conflict, any narration of events is only one among several competing narratives. Much ink has been spilled articulating these competing narratives; we have found Dowty, 2012, to be the most balanced and even-handed, rehearsing several competing narratives side-by-side and thereby acknowledging the truths they represent for various parties in the conflict.

This de facto state of war continued until 1979, when Israel and Egypt negotiated the first "land for peace" deal, in which Israel returned the Sinai to Egypt in exchange for recognition.[2] By 1982 the Israeli army had withdrawn from the Sinai and the border between the two countries opened at Taba, but it took another seven years for the border to be agreed upon at its current location east of Taba.[3] Despite the enthusiastic response of the Israeli public, who flocked to the Sinai and Egypt to visit heretofore inaccessible sites in the years after the treaty, the peace was and continues to be a "cold" peace: Very few Egyptians visit Israel, and Egypt still maintains restrictions on cultural and economic relations (Yaari, 2013).

Israel's second (and up until today, last) peace treaty with a neighbor was the peace accord signed with Jordan in 1994. Coming in the hopeful period following the signing of the Oslo Accords between Israel and the Palestinian Liberation Organization, the treaty between Israel and Jordan was signed by Prime Minister Yitzhak Rabin and King Hussein, in a ceremony taking place on the border between Eilat and Aqaba—the site at which the Arava Terminal would be established. As with the treaty with Egypt, many Israelis enthusiastically travelled to Jordanian sites like Amman, Aqaba, and Petra; on the other hand, fewer Jordanians travelled to Israel. As with Egypt, it was (and continued to be) largely a "cold" peace in which normal relations were curtailed by Jordan. However, given Jordan's significant Palestinian population, many Palestinians in Jordan travelled to visit family in the West Bank, East Jerusalem, or even parts of pre-1967 Israel. Palestinians in the West Bank and Gaza were granted permits to travel to or study in Jordan, but these depended on the level of cooperation or violence between the Palestinian Authority and Israel, as well as the "good will" of Israel. It is in part this back-drop—the long wait, from the Israeli perspective, for peace and acceptance by its neighbors and the relief and even exuberance when it arrived, however piecemeal; the ongoing tensions and outstanding issues, from the Jordanian and Egyptian perspectives, that remain to be resolved; and the domestic political considerations of how to handle a peace that is unpopular with a large segment of the population—that form the contexts for the border narratives of the land crossings between these countries.

---

2 Anwar Sadat's peace deal with Israel under Prime Minister Menachem Begin resulted in Egypt's ejection from the Arab League, a ban that held until 1989, when it was reinstated. Sadat himself was assassinated in 1981 by members of the Egyptian Islamic Jihad, in part due to his role in the peace deal. See Quandt, 1986.

3 During the period in which Israel held the Sinai, a resort hotel and casino was built in Taba and became a popular tourist destination. When the Israelis completed their withdrawal in 1982, they claimed that the actual border of the British Mandate (as well as the Ottoman sandjak) was west of Taba, and that the hotel was therefore in what should be Israeli territory. An international panel was convened, and decided that the border indeed ran east of Taba, and that the hotel was therefore in Egyptian territory. In the negotiations it was decided that Taba would have a special status, with visitors crossing the border and only going to the hotel being exempt from the border tax due upon crossing deeper into Sinai (Brinkley, 1989).

THE RABIN/WADI ARABA BORDER TERMINAL: COMPETING PERSPECTIVES ON "PEACE"

As discussed above, the Rabin/Arava border terminal, located just outside Aqaba (Jordan's only seaport) and the Israeli city of Eilat, was the site at which the peace treaty between these two countries was signed in 1994. Originally named the Arava Terminal in Hebrew, and the Wadi Araba Terminal in Arabic, Israel changed the name of their terminal in 2002 to the Yitzhak Rabin Border Terminal, in honor of the assassinated Israeli Prime Minister who signed both the Oslo accords with Yassir Arafat and the peace agreement with Jordan's King Hussein which maade the border crossing possible. These two names—one commemorating the leader who forged the peace agreement, the other labelling it geographically—could be said to reflect the different narratives presented in each terminal.

The Israeli terminal is replete with images and iconography commemorating Rabin and the peace between Israel and Jordan. When passing through the "departures" side of the terminal exiting Israel, there are only a few examples of that iconography (relative to the examples encountered in the arrivals side). As one passes the outdoor windows where passports are checked and taxes paid, one encounters a 3' × 3' mosaic depicting several human figures in various stances, over the words "CROSS ♥ BORDERS"; the caption identifies it as a 1996 work entitled "Austrian, Palestinian, Israeli, Jordanian," made by "the Bruno Kreisky Youth Peace Forum" of "Eilat, Aqaba, Bethlehem," and celebrates "a new generation crossing borders" (Figure 7.1). As one exits the terminal into the neutral zone between the countries, one sees a large blue billboard proclaiming "Peace שלום سلام" and bidding "Bon Voyage" in French, Hebrew and Arabic (Figure 7.2). These few direct images of peace are accompanied by a general air of informality, with whimsical cartoon-like figures pointing the way for travelers to follow, and a relative absence of visibly armed or militarily-uniformed guards.

As described above, this imagery is even more prominent as one passes through the "entering" terminal from Jordan: the same billboard is visible as one crosses the neutral territory, and the first space one moves into is the security screening room which simultaneously functions as a museum/memorial for Itzhak Rabin. While the room as a whole is dedicated to Rabin, the design of the memorial—the narrative constructed in the way one moves from the entrance of the room to its exit—carries one from his earliest years (on the left wall) and his family (on the right) to the peace agreement with Jordan. Passing through the security screening to reach the far wall reserved for photos of Rabin, King Hussein and Clinton in various poses, the peace agreement is figured for the traveler as the culminating achievement of Rabin's career.

Moreover, the peace agreement is not figured iconographically as an achievement of "*realpolitik*," or the formal cessation of hostilities; rather, it is presented as the outcome of a *relationship* between the two leaders, Itzhak Rabin and King Hussein. Photos depict them smiling together in various locations: looking over the Gulf of Aqaba; speaking together at what looks like an official function; standing with US President Bill Clinton. The most prominent photograph, and the one singled out by

7.1 "Cross ♥ Borders" mosaic, Yitzhak Rabin Border Terminal.

7.2 "Peace" billboard, exiting Yitzhak Rabin Border Terminal.

the terminal manager with whom we spoke as the most representative of the warm relations between the two, is a photograph in which the two men smoke together, King Hussein offering a light for Rabin's cigarette (see Figure 7.3). This emphasis on the relationship between these leaders is underscored by a memorial encountered after one leaves the security area and passes through passport control: emerging back into the open air and moving toward the exit, the traveler passes three olive trees arrayed alongside a polished limestone slab (Figure 7.4). The limestone marker reads "The Itzhak Rabin Border Terminal." Upon closer examination, each olive tree has a marker at its base: on the left, the plaque reads "In Memory of King Hussein Bin Talal;" in the middle, "In Memory of Itzhak Rabin;" on the right, "In Memory of Mrs. Leah Rabin." Although none of these now-deceased individuals is buried here, the elegiac effect is of grave markers, and—remarkably—Rabin is figuratively "buried" next to Hussein, as well as his wife Leah; all three are marked by the symbol of peace, the olive tree.

Thus the narrative constructed in the terminal's iconography and arrangement of space is one that foregrounds and celebrates the peace between Israel and Jordan, embodied and symbolized in the relationship between their two leaders. If this is the narrative that is being constructed, however, one might ask, for whom is it being constructed? Who is its audience? For it is noteworthy that this iconography is present, but relatively muted, in the exit terminal; it takes its most forceful form in

7.3 Yitzhak Rabin and King Hussein celebrating peace treaty, displayed in arrivals hall of Yitzhak Rabin Border Terminal. AP Photo 1994. © State of Israel.

the entrance terminal. Thus, we might say that the audience for this peace narrative is in particular those *entering* the country—international tourists and Jordanians, as opposed to Israelis. In fact, statistics show that of the 158,675 persons who crossed the border between January and May of 2013, 86,549 (or 55 percent) were tourists, and 39,258 (or 25 percent) were Jordanian workers who receive special visas to work in Israel. Only 21 percent, or 32,850 persons, crossing the border were Israelis; according to the manager of the Rabin Terminal, the vast majority of those are Palestinian Israelis, who see Aqaba as an inexpensive alternative to Eilat for a vacation on the Red Sea (IAA, 2013). Thus one might surmise that this narrative of peace is aimed primarily at an external audience, rather than an internal Israeli one. But to parse it even further, it is interesting that in the hall of photographs of Rabin, Hussein and the peace process, there is very little text or explanation. This makes the potential audience for this narrative even more specific: it assumes a degree of knowledge on the part of the viewer, such that they must already know who Rabin and Hussein are, or they wouldn't be able to "read" the narrative. Thus, the narrative, one might argue, is aimed at travelers crossing the border in general; among those travelers, it is particularly aimed at those who are entering the country from Jordan; and among those, it is particularly aimed at Jordanians crossing into Israel.

The iconography of the Jordanian terminal is quite different. Crossing from Israel, the traveler enters the Jordanian terminal through a gate under the royal gazes of both King Hussein and King Abdullah, as signs announce "Welcome to Jordan" in English and Arabic (Figure 7.5). Throughout the terminal, both entering

7.4 Olive Tree Memorial to King Hussein, Yitzhak Rabin and Leah Rabin. Yitzhak Rabin Border Terminal.

and exiting, the signage and iconography are almost exclusively of the royal family. Most often the pictures are of King Abdullah himself, in various contexts—in traditional dress, in his military officer's uniform, in a Western business suit—while sometimes he is next to his father, and even his son. When the traveler enters the small room housing Jordan's security X-ray screener, there are three portraits on the wall: King Hussein on the left; King Abdullah in the middle, and Abdullah's son, the Crown Prince Hussein.

In the Jordanian terminal, then, the narrative composed in space and imagery is of the presence and stability not just of the King, but of the royal line within which he is a link. But unlike in the Israeli terminal, there is no representation of either Kings' relationship with Israeli leaders, nor any representation of Israel whatsoever. There are no images of Rabin; there is no Hebrew visible; the only place where we saw Israeli and Jordanian flags in proximity to each other was on the baggage cart used to transport luggage across the neutral zone. Moreover, the audience for the images of the royal family might be said to be domestic: few of the images had any captions, and those that did were in Arabic, with no English.

Thus, examining the iconography and design of space on both the Israeli and the Jordanian sides of the Rabin/Wadi Araba Border Crossing, two contrasting narratives emerge. The Israeli terminal foregrounds and celebrates the peace process that made the crossing possible, and locates that peace in the relationship

7.5 "Welcome to Jordan." Entry to Wadi Araba Border Terminal, Jordan.

between Prime Minister Itzhak Rabin and King Hussein, who forged the peace agreement; this narrative is constructed most directly for an audience entering Israel through the terminal—international tourists and Jordanian citizens. The Jordanian terminal foregrounds the royal family—first and foremost the current King Abdullah, but also the line of succession that extends back to his father King Hussein and forward to his son Prince Hussein; this message of stability and continuity seems aimed at primarily a domestic audience.

This pattern is in fact repeated, in more muted tones, at the second border crossing between Israel and Jordan, the Jordan Valley/Sheikh Hussein border crossing near the Israeli town of Beit Sha'an, near the Sea of Galilee, opened in 1994 with the signing of the accords. As with the Rabin crossing, the decoration of the Israeli terminal centers on images of peacemaking. In the main hall through which both arriving and departing travelers pass, there is the same photograph of King Hussein and Prime Minister Rabin smoking cigarettes together. As with the Rabin crossing, this iconography is particularly directed at arriving travelers: passing through the entry gates where passports are checked and luggage scanned, travelers view a photographic montage of Rabin, Hussein, and US President Bill Clinton in the process of signing the agreements; above them is the Hebrew quote, translated as "all her paths are peace" (from Proverbs 3:16–18). On the Jordanian side, again, along with posters and photos of famous tourist sites in Jordan, there

7.6 Five generations of Hashemite kings, displayed in the arrivals hall of King Hussein Border Terminal, Jordan.

are images of King Abdullah, alone or in the company of his father, and indeed of his lineage going back several generations (an image similar to one displayed in the King Hussein Border Terminal (Figure 7.6)). There are no images referring to Israel or to the peace treaty with Israel.

Imagining the reasons for these different narratives is not difficult. For Israelis, the peace treaty and the opening of the border with Jordan was momentous, a hopeful moment breaking its regional isolation that had lasted 46 years, and perhaps raising hopes for peace with its remaining neighbors. King Hussein indeed garnered much admiration and affection from the Israeli public, first by being willing to make a peace agreement when other Arab countries wouldn't. Further, when in 1997 a Jordanian soldier opened fire on a group of Israeli schoolchildren who were on a fieldtrip to the "Island of Peace" that is located on the Jordan River between Israel and Jordan, killing seven and wounding six, King Hussein made the unprecedented move of visiting the parents of the dead to offer condolences. This show of empathy made him even more popular among Israelis. In foregrounding this peace to both Jordanians and international tourists crossing into Israel, Israel tells travelers (and themselves) the story of its peace efforts, and its willingness to keep working toward peace. On the other hand, the peace treaty and subsequent border openings is nothing particularly special from the Jordanian perspective: it has border crossings with all of its other neighbors as well. Moreover, the population of Jordan is itself approximately 50 percent Palestinian, including a large number of refugees from the wars in 1948 and 1967, and many of them are Jordanian citizens (Minority Rights, 2007). The relationship between the ruling Hashemites and the Palestinians, particularly the Palestinian Liberation Organization, has been tense and at times bloody (as we will elaborate below, with reference to the Allenby Bridge/King Hussein border crossing). Thus in terms of domestic politics, it makes perfect sense for the Hashemite regime to downplay its peace treaty with Israel, even at the border crossings into that country. Instead, given this potential inter-ethnic conflict, and increased challenges to Hashemite rule in the wake of the demonstrations for reform or regime change across the Middle East over the past several years, it is perhaps more important to the regime that the iconography proclaim the continuity and stability of the monarchy.

BORDER NARRATIVE AS PALIMPSEST: THE ALLENBY BRIDGE/KING HUSSEIN BORDER CROSSING

This dynamic, with the "border narrative" constructed at the Jordanian terminal aimed primarily at a domestic audience, and that at the Israeli terminal aimed primarily at an external (non-Israeli) audience, is repeated but with significant and revealing differences at the third land crossing between Israel and Jordan: the Allenby Bridge/King Hussein crossing between the West Bank near Jerusalem and Amman. This border crossing presents a uniquely complex "text" for analysis, due to its complex legal and diplomatic status as a border crossing for Israel, Jordan *and* the Palestinian Authority and the nascent state of Palestine. The border terminal is officially controlled by both the Palestinian Authority and Israel, but the Israel Airport Authority (IAA) administers the site, as it does all the other border terminals in Israel.

If, as scholars have argued, borders make material the construction of national identity—a process that in part takes place through the construction of the Other—by embodying a set of regulations designating who "belongs" inside, who belongs outside, and who may pass through, then the Allenby Bridge/King Hussein border crossing brings this process into high relief, and underscores the role of the passport in this legal and symbolic process. For the rules governing who can and cannot pass across this border are byzantine. As worked out in the Oslo Accords between Israel and the Palestinian Authority, this terminal is imagined as the eventual border crossing from Jordan into a Palestinian state, and thus now designated as the primary transit point for Palestinians travelling to and from Jordan, and for tourists who will eventually be coming to that state, as well as for pilgrims travelling to Jerusalem (the Muslim holy sites of which may be part of that eventual state). Thus those holding Israeli passports are not supposed to use this terminal to cross into Jordan: the Jordan River/Sheikh Hussein and the Rabin/Wadi Araba terminals described above are designated for Israeli passport holders (IAA, 2013a). Unlike the other two land crossings, visas are not granted at the terminal, but must be arranged at the Jordanian embassy in Tel Aviv—where Israeli passport holders who apply will be denied visas. On the other hand, Palestinians who live in East Jerusalem with permanent resident status are permitted to enter Jordan via this crossing,[4] as are Palestinian residents of the West Bank who hold either a Jordanian passport, a passport from the Palestinian Authority, or who have received a temporary Jordanian passport, *and* hold one of two "bridge cards" issued by the Jordanian government—a yellow card (indicating that their family moved to Jordan prior to 1983 and still has family in Jordan), or a green card (indicating that the person lived in the West Bank before 1983, and is only entitled to visit Jordan temporarily). Palestinians residing in Jordan who are Jordanian citizens also have a yellow card, which gives them access to their families in the West Bank without a visa (Research Directorate, 2004; US Consulate in Jerusalem, 2013).[5]

This confusing state of affairs is the result of the layers of competing and conflicting claims on the West Bank territory and its people among Israel, Jordan and the Palestinian Authority itself. For much of its modern history, Jordan has claimed the West Bank as part of its rightful territory, and the Palestinians as its citizens. From the Armistice in 1948 until the Six Day War in 1967, Jordan occupied

---

4 In the wake of the 1967 war, Israel held the 6.7 square km that had been called "East Jerusalem" (consisting of the Old City and Arab neighborhoods immediately surrounding the Old City), as well as the West Bank. While most of the West Bank has remained "occupied" or "disputed" territory, Israel quickly "annexed" what had traditionally been called East Jerusalem, as well as a broader area north, east and south of what had been East and West Jerusalem (approximately 64 square km), to form the larger municipality of Jerusalem, extending Israeli civil law and administration to these areas (as opposed to the military control that governs the West Bank). Palestinians living in this larger "East Jerusalem" (the term now is interchangeably applied to both the smaller and the larger area, depending on the context) were automatically made "permanent residents" of Israel, and eligible to apply for Israeli citizenship. Most Palestinian Jerusalemites refused to do so (a situation that continues to the present) and thus carry a blue identity card identifying them as residents of Jerusalem, rather than Israeli passports. As permanent residents, they have access to civil and municipal services that permanent residents anywhere in Israel do—the national health care system, education, municipal utilities, and so on—and can vote in municipal elections, but not national elections (BTselem, 2014). As with so many situations in the Israeli-Palestinian conflict, the precise legal status of East Jerusalem and its Palestinian residents—and the political ramifications of that status—is subject to debate and change. Some scholars have argued that Israel has not legally annexed East Jerusalem (for a variety of legal reasons and with varied political interpretations) (see Lustik, 1997); the precariousness of maintaining the requirements for "permanent residency" has led greater numbers of Palestinians in East Jerusalem to apply for, and receive, Israeli citizenship (see Barakat, 2012).

5 The intricacies of who has been, and is, eligible for Jordanian passports, and the relationship between Jordanian passports and residency and travel rights signified by the yellow and green bridge cards, is difficult to disentangle, and indeed subject to disagreement among knowledgeable sources: in response to the Canadian Immigrant and Refugee Board's query, Jordanian and Israeli human rights and refugee advocate groups had somewhat different understandings of the relationship between citizenship and who held yellow and green bridge cards. See Immigration and Refugee Board of Canada, 2004.

the West Bank and East Jerusalem (including the Old City with holy sites such as the Dome of the Rock and Al Aqsa Mosque), claiming this region that in the UN partition plan had been intended as the Palestinian state. Jordan officially annexed the West Bank and East Jerusalem in 1950 and 1953, respectively, declaring East Jerusalem the capital of the Hashemite Kingdom; this move was made palatable to other Arab states by claiming that Jordan was merely holding the land as a "trustee" until further settlement. For years official language included the name "Transjordan," used during the British Mandate to designate the territory on both sides of the Jordan River. During this period Palestinian residents on the West Bank, as well as thousands of refugees living in Jordan (in refugee camps as well as in cities like Amman) were given Jordanian passports and granted citizenship. After the Israeli takeover of the West Bank in 1967, the large number of Palestinians in Jordan, and especially the presence of armed forces of the Palestinian Liberation Organization under Yassir Arafat, posed a challenge to the Hashemite monarchy, and civil war broke out in 1970, culminating in the massacre of PLO forces and their expulsion from the country, known as Black September. By 1974, King Hussein came to recognize the PLO as the "sole legitimate representative of the Palestinian people"; by 1985, Jordan and the PLO were announcing plans envisioning a confederation between Jordan and a Palestinian state; and by 1988 Hussein had renounced any Jordanian claim on the West Bank (Kimmerling and Migdal, 1993). With this, and with the Oslo Accords in 1994 establishing the Palestinian Authority on the West Bank, Jordan began revoking the Jordanian passports of many of its Palestinian citizens, while West Bank Palestinians whose Jordanian passports expired were not allowed to renew them; they were granted temporary passports, or were issued Palestinian Authority identity papers with the Jordanian "bridge cards" signifying their ongoing special status vis-à-vis Jordan (Human Rights Watch, 2010).

Thus the terminal becomes the main point of transit for two specific kinds of travelers. First, it serves Palestinians from the West Bank and East Jerusalem and those living in Jordan who travel back and forth to visit family in these areas. Second, it is the primary avenue for pilgrims—Palestinians from the West Bank and East Jerusalem who travel to Jordan to take the haj, the traditional Muslim pilgrimage to Mecca and Medina or, going the other direction, for Muslims from around the Arab Middle East to travel to Jerusalem to visit the Dome of the Rock and the Al Aqsa Mosque. This is because West Bank residents are not permitted by the Israeli government to depart through Ben Gurion Airport. Conversely, Arab Muslims traveling to Jerusalem from countries that don't have diplomatic relations with Israel can only arrive in Amman, and then travel overland to Jerusalem. These two constituencies, we will argue below, profoundly shape the narratives on display at both the Israeli and the Jordanian terminals.

Thus in the tangle of regulations governing who can cross the border at the Allenby Bridge/King Hussein terminals and who can't, how long they can stay and under what visas, we see the echoes of competing claims of national identity and territorial sovereignty. Jordanian administrative procedures both acknowledge Palestinian nationality and independent territorial sovereignty, while *also* symbolically maintaining Jordan's own connection and claims. West Bank

Palestinians do not begin their traverse of the border at the Allenby Bridge/King Hussein terminal itself; instead they begin at a terminal in Jericho administered by the PA, implicitly recognizing the claims to national sovereignty of that government. Conversely, Jordanians, Palestinians and tourists who cross from Jordan to the West Bank and return across the border within 14 days, and do not leave the West Bank, do not have to pay for a new visa to reenter: as far as the Jordanians are concerned, in effect, that traveler has not left Jordan. At the same time, while officially administering the terminal in cooperation with the Palestinian Authority, Israel, in the form of the Israeli Airport Authority, effectively runs the terminal—a sign of Israel's continuing and for now decisive control over the territory. Thus, while these legal and diplomatic regulations do not create a "border narrative" in space and iconography in the way we have been discussing up to now, they do determine who is allowed or not allowed to access this space, and with what privileges, setting up who will be exposed to this spatial and iconographic narrative in the first place.

These competing claims and histories in part shape the border narratives constructed on the ground. Formally, national and ethnic identities, defined by the various forms of identification cards described above, are reinforced within the arrival and departure areas of both terminals, as signage separates travelers according to "tourists," "East Jerusalem" residents, and "Palestinians" (on the Israeli side), and "foreign passport" holders and "Jordanian" passport holders (on the Jordanian side). (There is no category of "Israeli" because Israeli passport holders are not supposed to use the terminal, although in practice they sometimes do.) In the arrangement of space, these designations blur, as all groups move through much the same space within the terminals. (It is important to note that Palestinian residents of the West Bank begin their traverse in the departure hall of the PA terminal in Jericho, a location whose design and iconography are beyond the scope of this study.) All travelers use the same departure hall in the Israeli terminal, for example, but travelers then must wait for buses to take them across the no-man's land of Allenby Bridge to the Jordanian terminal; there are different buses for foreign tourists on the one hand, and Palestinians (from East Jerusalem and the West Bank) on the other. Both groups are dropped off at the same terminal on the Jordanian side, but processed in different waves at different windows. Similarly, buses transporting all groups from the Jordanian terminal to the Israeli terminal drop everyone off in the same location for baggage pickup and security screening, after which signage designates different lines for "tourists," "East Jerusalem" and "Palestinians," but in practice all use the same entry hall (in fact, on the day we last visited the border, there was effectively one line into the entry hall, and distinctions between the groups materialized when having to pass through different booths for passport control). Beyond this check, tourists and East Jerusalemites proceed through common spaces to public transportation areas, where buses and taxis will take both groups to East Jerusalem. Palestinians heading for the West Bank are ushered to other spaces for transportation to the Jericho terminal.

Iconographically, the Jordanian and Israeli terminals compose narratives similar to the other two land crossings, with interesting differences in emphasis. In the Jordanian terminal, images of King Abdullah, his father the late King Hussein, and other members of the royal family predominate, along with posters promoting popular tourist destinations in Jordan. There are posters of King Abdullah similar to those at the other two crossings. Several images focus on the King himself seen in various roles and forms of dress (e.g., Western, traditional, military, etc.). Some emphasize King Abdullah's place in the royal lineage: one image places him in the line of successive Hashemite kings stretching back several generations (Figure 7.6); others place his portrait next to that of his father, or between his father and his son the Crown Prince. In this terminal, however, there are relatively more pictures of the royal family itself than in the others, including Queen Raina al Abdullah and their four children: there are several photos of King Abdullah and Queen Raina, individually, meeting with what look like "ordinary people"; there are also several family portraits, with the King and Queen surrounded by their four children in informal poses, evoking a domestic family scene. The most striking of these images is a family portrait found in the plaza just outside the arrivals terminal, where travelers gather to catch a bus or taxi into Amman (a plaza that also serves as access to the departures hall on the other side): on the left side, the royal family sits on a bench in a living-room like setting, dressed in casual attire; on the right, on a black background, an image of the map of Jordan, composed of three interlocking hands, each rendered in one of the remaining three colors of the Jordanian flag (red, white and green), with the caption underneath stating "We are all Jordan" (Figure 7.7).

In this poster, the narrative of the continuity and stability of the royal family takes on additional significance, and one that maps neatly onto this particular terminal as the one primarily designated for Palestinians. In the interlocked hands, the poster conveys an image of an inclusive Jordanian national identity, one that links Hashemites with Palestinians and with other ethnic and tribal groups that make up the country's population (most notably, Bedouin who make up approximately 33 percent of Jordan's population [Minority Rights Group International]). This narrative of Palestinian and Hashemite unity under the Jordanian monarchy is doubled by the royal family itself: Queen Raina is herself Palestinian. Thus the prevalence of images of Queen Raina at this terminal seems aimed at the particular constituency passing through its gates, giving a particular iconographic inflection to the broader imagery narrating the stability of the monarchy seen at the other land crossings.

If the iconography and symbolism of the King Hussein terminal present the same general narrative we argue is constructed at the other two land crossing, and tailor it to the particular travelers imagined passing through here, then the same might be said of the Israeli narrative constructed in its terminal, where the narrative of peace that we suggest is foregrounded elsewhere takes a form specific to the audience imagined here. Passing through the departures terminal, very little signage or iconography of any kind is present. In the arrivals hall, however, the picture is quite different. Immediately upon disembarking from the bus bringing passengers over the Allenby Bridge from Jordan, travelers pass a quick document checkpoint, and move into a larger enclosed room with airport-like security scanners. One wall of this space presents a mural depicting silhouetted figures meeting on a bridge (one woman, five men, all somewhat incongruously dressed in Western suits and tuxedos), shaking hands, under a sign proclaiming "Welcome"

7.7 "We Are All Jordan." Billboard displayed outside King Hussein Border Terminal, Jordan.

in English, Hebrew and Arabic (Figure 7.8). After passing the screening machine, travelers go into a larger room, walking under a sign designating three lines for three types of identification—"tourist," "East Jerusalem," and "Palestinian"—but on the day we visited, there was just one line through which everyone entered the room. Inside there is a line of booths for passport control forming one wall-like block dividing the space; the three remaining walls were covered with a number of posters or photographs, each depicting a scene from the Old City of Jerusalem, most frequently a view of the Al Aqsa Mosque/Dome of the Rock. After going through passport control, tourists and East Jerusalem Palestinians continue through another set of checks and into one space, where the photographs and posters shift to general tourist sites and nature scenes from around Israel proper. Palestinians destined for the West Bank enter different lines headed for another space (on their way to Jericho).

In this array of iconography, several things emerge. First, as at the Rabin crossing, the terminal foregrounds the peace process and peacemaking with the mural of the bridge. As there is no such iconography in the departures hall, the audience for this narrative is those who are imagined to be entering through this terminal—Palestinians returning from Jordan, Jordanians visiting the West Bank, and pilgrims headed to Jerusalem. (One might surmise that, given the iconography of the "bridge" as a symbol of peace, the peace process indexed here is the peace between

7.8 "Welcome" mural, arrivals hall of Allenby Bridge Border Terminal, the West Bank.

Israel and Jordan, not necessarily that between Israel and the Palestinian Authority, perhaps because there has been no final peace agreement yet between Israel and the PA.) Second, and perhaps more remarkably, the distinct imagery displayed in the passport control room indicates that the narrative at Israeli border terminals *varies* from terminal to terminal, to speak specifically to the constituencies that the IAA (either personnel in the general headquarters, or individuals "on the ground" at the sites themselves) imagine will be crossing there, rather than being the result of a standard policy centrally administered to apply to all border terminals. Thus the iconography of the Al Aqsa mosque and the Old City of Jerusalem serves as an explicit acknowledgement, and even invitation, to the many pilgrims Israel knows are coming through this terminal. The final opening up in the last space to general tourist images of Israel might represent the steps Israel hopes those pilgrims might make—from visiting the holy sites in the Old City of Jerusalem, to perhaps visiting other sites in the country.

Thus, looking at all three land crossings between Israel and Jordan, several patterns emerge. At each terminal, the arrangement of space and iconography combine to construct border narratives, a first opportunity (or last chance) to present the traveler with a particular message that constructs the national space they are about to enter (or leave), a relation to the adjacent space, and a position for the traveler in that narrative. For the Jordanian terminals, the iconography centers almost exclusively on images of the King, often with his father or his son, aimed primarily at a domestic audience. This iconography and the narrative of royal stability that it conveys varies little across the three terminals, with one exception: at the King Hussein crossing, which sees the most Palestinian traffic, the iconography emphasizes the royal family, including Queen Raina, who is Palestinian. These pictures fit other iconography only found at this terminal, evoking national solidarity between Palestinians and Jordanians of other backgrounds. On the other hand, Israeli terminals consistently take as their audience travelers *entering* the country, that is, not Israelis. The narratives aimed at such travelers celebrate the peace between the two countries in various ways. The particular form this celebration of peace or welcoming takes, however, varies according to the main groups of imagined travelers traversing that particular border. While the Rabin crossing emphasizes—primarily to an external audience of Jordanians and tourists entering Israel at that crossing—the peace between Israel and Jordan as personified by the relationship between Rabin and King Hussein, at the Allenby Bridge Terminal the message aimed at this external audience—here Palestinians and pilgrims from around the Arab world—is one of generic "bridges of peace," and a recognition of their particular interest in Jerusalem as a holy Islamic site. This variation-within-consistency, however, does not seem to be the result of a centrally planned policy on the part of the IAA, guiding the use of iconography and artwork at each terminal. Interviews with IAA personnel, including both duty managers on the ground and the Manager of Public Affairs Irit Levy, suggest that at least some of these decisions on posters and artwork were ad hoc, and not coordinated—in fact, some thought the research revealed something that they should, in the future, coordinate in a more mindful way (Levy 2013). In fact, there is evidence that the border narratives constructed in the Israeli terminals—tailored to address and welcome visitors imagined to be crossing at each location—might in some ways be generated by the status and history of the Israeli Airports Authority as a public corporation that in 1977 was tasked with running Israel's airports; only in 1980, with the opening of some of the land borders discussed here, was the IAA tasked with running those terminals as well. This might account for the airport-like feel of the terminals—well lit, open, decorated—and more importantly, for a culture of customer orientation designed to appeal to and attract travelers in a way uncharacteristic of government bureaucracies. This ad hoc customer orientation, and its possible tensions with the security apparatus also embodied in a border terminal, is central to the narrative of the Taba border crossing between Israel and Egypt.

NOSTALGIC PAST MEETS TENSE PRESENT: TABA BORDER CROSSING BETWEEN ISRAEL AND EGYPT

Passing through the Israeli arrivals and departures terminals at the Taba border crossing presents a contrast to the other borders in this study. There is no foregrounding or celebration of the peace process or peacemaking nor any commemoration of the two leaders who made it possible, Israeli Prime Minister Menachim Begin and Egyptian President Anwar Sadat. Rather, it resembles a border between any two states; perhaps the contrast between the 34 years of peace between Egypt and Israel and the 19-year-long peace between Jordan and Israel accounts for the difference. In fact, in the Israeli terminal there are no images of political leaders, past or present, as is common at so many other international borders. What is evident, however, are separate narratives in the arrivals and departures halls that seem to be aimed at different audiences—tourists arriving to vacation in Eilat and Israelis exiting to visit Egypt.

In the arrivals hall, the narrative is relatively straightforward. As travelers come into Israel from the Egyptian side, there are no overtly political statements regarding peace, but as one passes through the security screening area one encounters colorful murals of the beautiful fish and corals of the Red Sea, a nod to one of the primary tourist attractions of Eilat that presumably draws tourists across the border. Reinforcing this, the terminal includes an open-air balcony looking out at the Red Sea, where one can stop and enjoy the view. The narrative of a vacation destination continues in the row of photographs lining one hall, showing the smiling images of the diverse employees of the terminal, each with a quote about the importance of smiling. These "customer service" messages—which one might argue are aimed as much at the employees themselves as the tourists traveling through, since they are entirely in Hebrew—are reinforced with the presence of a customer feedback form and drop-box.

If the arrivals hall constructs a narrative of resort vacations aimed at tourists, the departures hall constructs a narrative primarily for Israelis. Here, the traveler moves through passport control into a hallway at the end of which is another ID check; the left side of the hall is lined with the same gallery of smiling employee portraits described above. On the right hand wall, however, are arrayed a series of framed photographs, each showing an image of the border terminals, such as they were, in the early 1980s when the border opened to travel for Egyptians, Israelis and tourists. Rather than the kind of serious celebration of peace that one finds at the Rabin crossing, these photos convey a nostalgic, almost comic glance at the humble beginnings of the border crossing, at perhaps a more innocent and hopeful time. The images, for example, show a small caravan that was the first terminal on the Israeli side; another shows bearded, long haired hippies walking their luggage across in carts between the two terminals; a third shows several men in swimming trunks standing outside a checkpoint (a one-man portable module) before the Egyptian terminal (a one-room booth) (Figure 7.9). In the photos, virtually no military presence is visible, which is surprising given that Egypt was Israel's strongest military rival. What the photographs do depict is the growth of the terminal itself over the years.

The audience for this humble and humorous glance back at the history of the terminal, we suggest, might be imagined as tourists of all nationalities rather than an Egyptian audience, because there is no such imagery in the arrivals hall and because few Egyptians cross the border here other than for work. More specifically, it seems, it is for an *Israeli* audience: the captions are all hand-typed in Hebrew, with

7.9 Photograph displayed in exit hall, Taba Border Terminal, Israel.

no English or Arabic translations. And this humorous memorial is not, as discussed above, the result of a central decision made in Tel Aviv; according to the Israeli duty manager, it is something that the employees of the terminal put together themselves with photos taken by the then-manager of the crossing. And there are other elements that suggest the signage and iconography are directed primarily at Israelis: here, and at no other terminal, there is a sign reminding (Israeli) travelers, in Hebrew, that "the face of the country is you! Represent us with honor." We were unable to ascertain the occasion for this admonition about respectful behavior abroad but again, it is a message clearly directed to a domestic audience.

While the narratives of the Israeli terminal are shaped by signage and wall art, the narrative of the Egyptian terminal takes shape as much through its architecture as its wall art. While the Israeli photographs described above nostalgically remember the border terminals' humble beginnings, no such nostalgia is evident here. If there is nostalgia, it is for a very different past, one of grandeur rather than humble beginnings. The Egyptian terminal, unlike any of the other terminals researched for this project, has an architecture that aspires to grandeur: the traveler ascends several stairs to enter a central three-story entryway/atrium, the front wall of which is all glass (Figure 7.10). (All other land terminals are one story, or even, at the Rabin/Wadi Araba terminals, long narrow buildings, rather like double wide trailers, end to end, with service windows facing the outside where travelers stand.)

7.10 Arrivals hall, Taba Border Terminal, Egypt.

Flanking the entryway are two-story high arcades, supported by rows of white pillars, evoking perhaps the grand pillars of ancient Egyptian sites like Karnak. The arrivals and departures terminals extend back from the entryway, on the left and the right; each hall is three stories high from floor to ceiling. Travelers proceed from station to station (security screening, passport control, departure tax, etc.) down the center of the hallway (with side trips to various bureaucratic offices located down the middle of the building, giving them access to both arrival and departure halls). In niches high up where wall and ceiling meet, there are pictures of ancient Egyptian art and hieroglyphics.

The narrative of this terminal, then, links the modern state of Egypt with the grand civilization of ancient Egypt. At the same time, iconography directly related to the modern state was curiously absent at the time of our research: one could see at intervals along the wall blank spaces where large posters had recently been; upon inquiry, terminal employees replied evasively and indirectly, indicating that these had been portraits of Egyptian presidents. While it is likely that in the months prior to our visit there were pictures of (first) former Presidents Hosni Mubarak and (later) Mohamed Morsi, we crossed in July 2013, right after the ousting of Morsi by the Egyptian military. It was a transitional period, during which photographs of the new president had not yet been installed.

The festive narrative of resort tourism at the Israeli arrivals terminal, the nostalgic celebration of the growth of the terminal at the Israeli departure terminal, and the proud evocation of Egyptian glory, are all in tension however with the evidence on both sides of the border of the increasingly tense security situation there, and in the Sinai more broadly. Since the collapse of the Mubarak government, Bedouin smugglers and extremists linked to al Quada have taken advantage of Cairo's weakening hold on the Sinai to pursue increased drug smuggling, human trafficking of refugees and migrants from West Africa trying to make it to Israel, kidnapping, and using the border areas as bases to launch attacks on both Egyptian forces and into Israel (Katz, 2011; Yagna, 2012; Lappin, 2013; Laub, 2013). Consequently, security was some of the tightest we had seen at any of the borders. There seemed to be more visible security on the Egyptian side of the border than the Israeli side, with multiple layers of security between the Egyptian arrivals terminal and the Taba Hilton. As one enters the Israeli side of the border, the employee checking passports was behind thick plexiglass in a hardened booth, with one slot, about 6 inches by 2 inches, through which one passes the passport—the most fortified space we saw in any of the borders. The security search in the Israeli arrivals hall was the most thorough we encountered at any location during 2 months in Israel, including at Ben Gurion Airport. The heightened tension at this border crossing reflects the heightened security concerns all along the Israeli border with Sinai.

CONCLUSION: BORDER TERMINALS AS NARRATIVES OF PEACE AND CONFLICT

Examining the border terminals between Israel and Jordan, and Israel and Egypt, it is clear that borders function not only as "boundaries of exclusion and inclusion" (Newman and Paasi, p. 194)—as lines that determine who is inside, who is outside, and who can pass through—but as narrative spaces in which meaning is constructed in the process of passing through. The arrangement of space, the use of images, artwork, and signage compose stories that position travelers in relation to the national spaces they are about to leave or enter, constructing a narrative as to who they might be, why they might be there, and their relation to that national space. In the Israeli and Jordanian cases studied here, those narratives fell into a particular pattern: in the Israeli terminals, the narrative highlights the peace processes that have only recently (relative to the country's history) made such border crossings possible; the audience for this narrative, moreover, seems to be the foreign tourists, and especially Jordanians, imagined to be crossing at these points. The Jordanian terminals, on the other hand, contain no such references; instead, the political symbolism of the Jordanian terminals seemed aimed at a domestic audience, and served to underscore the stability and continuity of the monarchy. Given this relatively consistent narrative structure across the three border terminals examined, interesting variations within this structure occurred at each terminal, especially in the Israeli terminals. The iconography at the Jordanian terminal at the King Hussein crossing, where Palestinians living in Jordan and Palestinians in the West Bank and East Jerusalem are the major constituency, emphasized for example the idea of national unity across different ethnic groups in a way that was not evident in other terminals, including foregrounding the royal family of the Hashemite King Abdullah and the Palestinian Queen Raina. The Israeli terminals also exhibited varied artwork and signage that seemed to address and welcome the specific groups that were imagined to be crossing at that location. How these narratives are composed remains an interesting question: in the Israeli cases, the tailoring of these narratives seems not to have been the result of a thought-out program codified and disseminated from a central administration, but rather a relatively ad-hoc process reflecting the ethos of the managers and designers of the individual crossings. At the Israeli-Egyptian border, a different set of narratives emerged: the peace process was celebrated by neither the Israelis nor the Egyptians; rather, the Israeli terminals constructed narratives priming arriving tourists for a resort vacation (in the arrival hall), or reminding exiting Israelis of the history of the terminal in a way expressing both nostalgia and progress. The Egyptian terminal also constructs a narrative of the past, connecting modern Egypt to the grandeur of ancient Egypt.

More broadly, these sites illustrate the ways in which border terminals are spaces that construct meaning through the dynamic interaction between the symbolism, iconography and the arrangement of space that present a border narrative and the political tensions and conflicts, in the form of security concerns, legal and diplomatic restrictions, and so on, that impinge on that narrative. All the sites visited exhibited the tensions between the historical narratives they seemed to want to construct, and the pressures of contemporary security concerns on these border narratives. What emerge are spaces that are "bordering on peace:" marking (either exuberantly or mutedly) the peace that has been achieved, and marked by the difficulties that remain in the way of a true peace.

ACKNOWLEDGMENTS

This research was made possible in part by a Michigan State University Office of Study Abroad Travel Extension Grant, as well as funding from the Residential College in the Arts and Humanities and James Madison College of Michigan State University. We would like to thank the several station managers of the Israeli border terminals who made themselves available for interviews for this article, but who must remain anonymous without the approval of the IAA. We also would like to thank Professor Saliba Sassar, Professor of Political Science and Associate Vice President for Global Initiatives at Monmouth University, for his assistance in Arabic translations.

REFERENCES

Abdullah. 2011. *Our Last Best Chance: The Pursuit of Peace in a Time of Peril*. New York: Penguin.

Amnesty International. 2013. "Egypt/Sudan: Refugees and asylum seekers face brutal treatment, kidnapping for ransom, and human trafficking." London: Amnesty International Publications [online]. Available at: https://www.amnesty.org/en/documents/afr04/001/2013/he/.

Barakat, R. 2012. "Quietly, East Jerusalem Palestinians acquiring Israeli citizenship." *+972 Magazine*, 20 May [online]. Available at: http://972mag.com/quietly-east-jerusalem-palestinians-are-becoming-israeli-citizens/46298/ (accessed 18 January 2014).

Barzilay, E., Enzel, Y., and Amit, R. 2000. "Constructing synthetic time series of rainfall events for a hyperarid environment, southern Arava, Israel." In M. Hassan, O. Slaymaker, and S.M. Berkowicz (eds) 2000. *The Hydrology-Geomorphology Interface: Rainfall, Floods, Sedimentation, Land Use*. Wallingford, UK: IAHS Press, pp. 29–42.

BBC News. 2001. "Profile: Jordan's Queen Rania," 7 November [online]. Available at: http://news.bbc.co.uk/2/hi/middle_east/1632614.stm (accessed 18 January 2014).

Brinkley, J. 1989. "Signing of agreement turns over last of Sinai to Egypt." *New York Times*, 27 February [online]. Available at: http://www.nytimes.com/1989/02/27/world/signing-of-agreement-with-israel-turns-over-last-of-sinai-to-egypt.html (accessed 16 January 2014).

B'Tselem. 2014. "Legal status of East Jerusalem and its residents" [online]. Available at: http://www.btselem.org/jerusalem/legal_status (accessed 18 January 2014).

Dowty, A. 2012. *Israel/Palestine*. New York: Polity.

Immigration and Refugee Board of Canada. 2004. "Jordan: Jordan's treatment of failed refugee claimants who are returned to Jordan or persons who have exited the country illegally or whose permission to leave has expired; whether there is a distinction made between citizens of Jordan, stateless Palestinians from the Occupied Territories, and stateless Palestinians who reside in Jordan under UNRWA registration; possibility of torture or the existence of a risk to life or a risk of cruel and unusual treatment or punishment upon return" [online]. Available at: http://www.refworld.org/docid/41501c2623.html (accessed 18 January 2014).

Israel Airport Authority. 2013. *Misofi-Gvul vSadeh Tufah Pnai-Artzim* [PDF].

Katz. Y. 2011. "Eight killed in massive terror assault near eilat." *The Jerusalem Post*, 19 August [online]. Available at: http://www.jpost.com/Defense/Eight-killed-in-massive-terror-assault-near-Eilat (accessed 17 January 2014).

Kimmerling, B. and Migdal, J.S. 1993. *Palestinians: The Making of a People*. New York: Free Press.

Lappin, Y. 2013. "Sinai terrorists fire rocket at Eilat, Iron Dome intercept." *The Jerusalem Post*, 12 August [online]. Available at: http://www.jpost.com/Defense/Eilat-hit-by-rocket-from-Sinai-as-sirens-blasts-heard-around-city-322859 (accessed 18 January 2014).

Laub, Z. 2013. "Egypt's Sinai Peninsula and security." Washington, DC: Council on Foreign Relations [online]. Available at: http://www.cfr.org/egypt/egypts-sinai-peninsula-security/p32055 (accessed January 2014).

Levy, I. 2013. "Discussions on Israel Airport Authority policies at the border terminals with IAA Manager of Public Affairs" [email] (Personal communication, July 24, August 12, December 22).

Lustik, I. 1997. "Has Israel annexed East Jerusalem?" *Middle East Policy* 5(1), 34–45.

Mekonnen, D.R. and Estefanos, M. 2001. "From Sawa to the Sinai Desert: The Eritrean tragedy of human trafficking," Social Science Research Network [online]. Available at: http://ssrn.com/abstract=2055303 or http://dx.doi.org/10.2139/ssrn.2055303.

Minority Rights Group International. 2007. "World Directory of Minorities and Indigenous Peoples—Jordan: Overview" [online]. Available at: http://www.refworld.org/docid/4954ce4ec.html [accessed 18 January 2014].

Newman, D. 2011. "Contemporary research agendas in border studies: An overview." In Wastl-Walter, D. (ed.) 2011. *Ashgate Research Companion to Border Studies*. Farnham: Ashgate Publishing, pp. 33–49.

Newman, D. and Paasi, A. 1998. "Fences and neighbors in the postmodern world: Boundary narratives in political geography." *Progress in Human Geography* 22(2), 186–207.

Quandt, W. 1986. *Camp David: Politics and Peacemaking*. Washington, DC: Brookings Institute.

Sack, R.D. 1986. *Human Territoriality: Its Theory and History*. Cambridge: Cambridge University Press.

Shapiro M.J. and Alker, H. (eds) 1996. *Challenging Boundaries: Global Flows, Territorial Identities*. Minneapolis, MN: University of Minneapolis Press.

Taylor, P.J. 1994. "The state as container: Territoriality in the modern world-system." *Progress in Human Geography* 18, 151–62.

US Consulate General Jerusalem. 2013. "Jordan's Border Crossings with the West Bank and Israel" [online]. Available at http://jerusalem.usconsulate.gov/pdfs/new-acs-website---crossings-west-bank-israel-jordan.pdf.

Yaari, E. 2013. *Israel-Egyptian Peace: Forty Years After the 1973 War and Still Holding*. Washington, DC: The Washington Institute.

Yagna, Y. and Cohen, G. 2012. "IDF soldier killed, another wounded in clashes on Israel-Egypt border." *Ha'aretz*, 21 September [online]. Available at: http://www.haaretz.com/news/diplomacy-defense/idf-soldier-killed-another-wounded-in-clashes-on-israel-egypt-border-1.466126 (accessed 18 January 2014).

Yaron, H., Hashimshony-Yaffe, N., and Campbell, J. 2013. "Infiltrators or refugees? An analysis of Israel's policy towards African asylum-seekers." *International Migration* 51, 144–57.

8

The View from Above: Reading Reunified Berlin

Julia Walker

The view from above, however one may find it described—as the aerial perspective, the bird's eye view, the panorama—is always at once totalizing and incomplete. One might think of writer Guy de Maupassant coping with the incursion of the Eiffel Tower into his beloved Paris, as recounted by Roland Barthes. According to Barthes (1964, p. 236), "Maupassant often lunched at the restaurant in the tower, though he didn't care much for the food: *it's the only place in Paris*, he used to say, *where I don't have to see it*." Maupassant's acid remark is also characteristically shrewd, revealing a mode of vision typical of modernity in which the plenitude of the view from above is offered up as evidence of the viewer's apparent *mastery* of the realm below. In fact, Barthes' meditations on the Eiffel Tower, in his essay of the same name, were hardly limited to that single monument; rather, Barthes had in mind that they could be applied to "all panoramic vision," in which the viewer's perspective (always mediated by architecture) represents "the only blind point of the total optical system of which it is the center" (p. 237). Barthes' analysis of the Eiffel Tower is indispensable for understanding the visual regimes of modernity in general, but, as a way of interpreting historical as well as visual landscapes, it offers particular insight into the fragmented urban plans of late modernity. One such plan is Axel Schultes and Charlotte Frank's master plan for the government district of Berlin, a scheme that, perhaps even more than its 1992 competition rivals, asks its viewer to consider the bird's-eye view of paramount significance. For it is only in the view from above that Berlin's boundaries, borderlands, and still-fallow frontiers can be legibly assembled.

Contemporary Berlin is a city whose recent architectural interventions demand to be "read" as parts of a narrative, one that includes urban forms built, destroyed, and never realized and that includes histories that occurred, that were hoped for, and that are still imagined to be possible. This element of fantasy is unsurprising; as Barthes reminds us, "architecture is always dream and function, expression of a utopia and instrument of a convenience" (p. 239). In viewing the city, we have become overfamiliar with the panoramic view, in which the panorama is "an image we attempt to decipher, in which we try to recognize known sites, to identify landmarks" (p. 243). In fact, this process of deciphering is key to Barthes' theory of the semiotic function of the Eiffel Tower, because, as he argues, perception of it is necessarily dialectical. Characterized by an initial "euphoric vision" at how "continuous" the landscape appears from "the bliss of altitude" (pp. 243–4), the viewer then must analyze this image. By locating recognizable elements within the image, by performing "the incomparable power of *intellection*" (p. 242), the viewer is reoriented and reconnected with the landscape. To become meaningful, Barthes' panorama demands a readerly intervention.

For Barthes, "to perceive ... from above is infallibly to imagine a history," in which "the mind finds itself dreaming of the mutation of the landscape which it finds before its eyes" (p. 244). In other words, the panoramic view offers a depth of field that spans time as well as space, creating the same "continuousness" of historical development as that created in the visual landscape by the aerial perspective. Yet this historical depth is artificial and subjective, the product of what Barthes terms "a kind of spontaneous anamnesis" (p. 244) that allows the viewer to perceive and organize history as orderly strata that accumulate into the present. Thus, the aerial perspective, whether deployed from the top of a tower or from the gaze at a master plan, permits the viewer to see more clearly the irruptions of the past into the present. Yet this apparently complete view of history is, in fact, a pile of fragments that equal a whole only through the intellectual process of viewing from above.

Despite its official reunification, Berlin has maintained its image as a site of fragmentation and ongoing conflict, and the city's status as notionally if no longer actually divided has posed particular difficulties and opportunities for the architects and planners who have been charged with the task of representing the capital of the new Germany. During the process of reunification, perhaps no area of the city generated as much anxious discussion as the Spreebogen (Figure 8.1), the bend in the river Spree that defined the course of the Wall and that embraced the past and future home of the federal government. Occupying a parcel of about 150 acres, the site lies at the western edge of the historical city center, north of the Brandenburg Gate and the Tiergarten and slightly northwest of Unter den Linden, Berlin's historical main thoroughfare. Axel Schultes, the architect whose firm was responsible for master planning this government district, claimed that the challenge posed by this terrain was how "to coax the soul out of the Spreebogen, the genius loci, to pour its historical and spatial dimensions into the mold of a new architectural allegory" (quoted in Zwoch, 1993, p. 49). Schultes and his partner, Charlotte Frank, proposed that the new plan take the form of a "Band des Bundes" (or "ribbon of federal buildings") spanning the Spree twice and crossing the former boundary between east and west; this precinct of rupture and conflict would thus be transformed into the very image of Germany's new national self-confidence (Figure 8.2). But by what means could that transformation take place? Considering the words of the architects proves revealing. Specifically, Schultes' reflection that the Spreebogen's vexed history must be refigured as a "new architectural allegory," is often cited but seldom examined. What does the architect intend by the use of the term "allegory?" Does he merely mean that his plan should embody, express,

8.1 Aerial view of Spreebogen site, 2008. Deutsch: Blick auf den Berliner Hauptbahnhof vom Helikopter aus. Photograph by Benjamin Janecke. Reproduced under Wilkicommons Licence, http://commons.wikimedia.org/wiki/File: Berlin Hauptbahnhof.jpg.

or even enact reunification, or is the reference more particularized than that? In fact, Schultes' words are not incidental, nor does he merely seek to lend literary theoretical ballast to the visual art of urban planning. Rather, allegory is the key formal device governing his master plan. Even more significantly, it is the mode in which the plan of the Spreebogen discloses its historical and political past, as its meaning is continuously deferred or absent.

The notion of allegory addresses questions of cohesion, of the conflicting impulses to come together or unify and the opposite, to disperse. Indeed, what Schultes and Frank would have found when they began to research the site was that political unification was a fantasy that had haunted the Spreebogen throughout modernity. In the early nineteenth century, the area served as drilling grounds for the Prussian cavalry. Towards the end of the century, it became the home of the Reichstag (or parliament) building after the first unification of Germany. The antagonism between the dynastic Hohenzollerns and the newly formed parliament, or "the imperial monkey house," in Wilhelm II's terms, meant that parliamentarians spent their days far from the royal palace, stranded in what was then known as the "Sahara Desert" (Koshar, 1993, p. 62). This spatial disintegration of government functions meant that, in contrast to other European capital cities, Berlin lacked a clear, central area for government use. Planners sought throughout the twentieth century to find ways of uniting these distant spaces, and competitions were

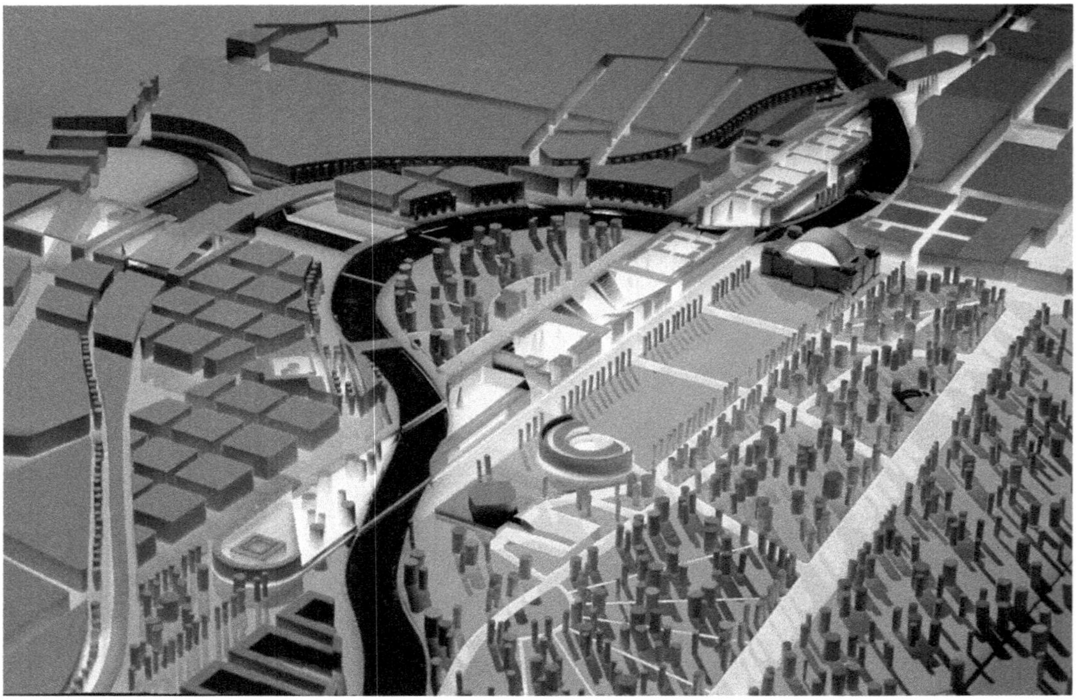

8.2 Axel Schultes with Charlotte Frank, *Band des Bundes*, 1992. Project for Spreebogen, Berlin, Germany. All plans, drawings and texts are subject to the authorship and creatorship of Axel Schultes and Charlotte Frank.

regularly held in the late nineteenth and early twentieth centuries to solve the seemingly intractable problem of the marginalization of elected representatives. From 1908 to 1911, a series of competitions were specifically aimed at balancing the urban emphasis between the site of the royal palace and the new government area, with results that many found only somewhat satisfactory.

Flush with republican zeal after the abdication of the Kaiser, Weimar architects sought to express the area's new political centrality. In Hugo Häring's 1927 proposal (Figure 8.3), the Platz der Republik in front of the Reichstag was to be surrounded by grandstands for public events, turning the former imperial square into a "Forum of the Republic" (quoted in Sonne, 2006, p. 199). Häring clustered government functions within high-rise buildings dispersed around the square, creating a modernist office complex reminiscent of the administration center in Le Corbusier's 1922 project for a Contemporary City for Three Million Inhabitants. However, whereas Le Corbusier's project married highly rational city planning with an almost picturesque emphasis on siting and green space, Häring's plan evinced a strong axial focus and a new, emphatic north-south boulevard. His intention was to undermine the imperial city planning that currently existed; his new north-south axis, cutting across Unter den Linden, would serve as "a distinct and clear line through this axis of the rulers" (quoted in Sonne, 2006, p. 199). Hans Poelzig's project of 1929 similarly placed government offices in towers that radiated into the curve of the Spree and visually lightened the bulky Reichstag. Like many Weimar architects, Häring and Poelzig saw the language of the efficient, clean-lined tower as one that was inherently democratic.

8.3 Hugo Häring, project for Spreebogen, 1927. Hugo Häring © 2014 Artists Rights Society (ARS), New York/VG Bild-Kunst, Bonn.

8.4 Albert Speer, model for "Welthauptstadt Germania," 1939. Bundesarchiv, Bild 146III-373/CC-BY-SA. Reproduced by permission of Das Bundesarchiv, Germany.

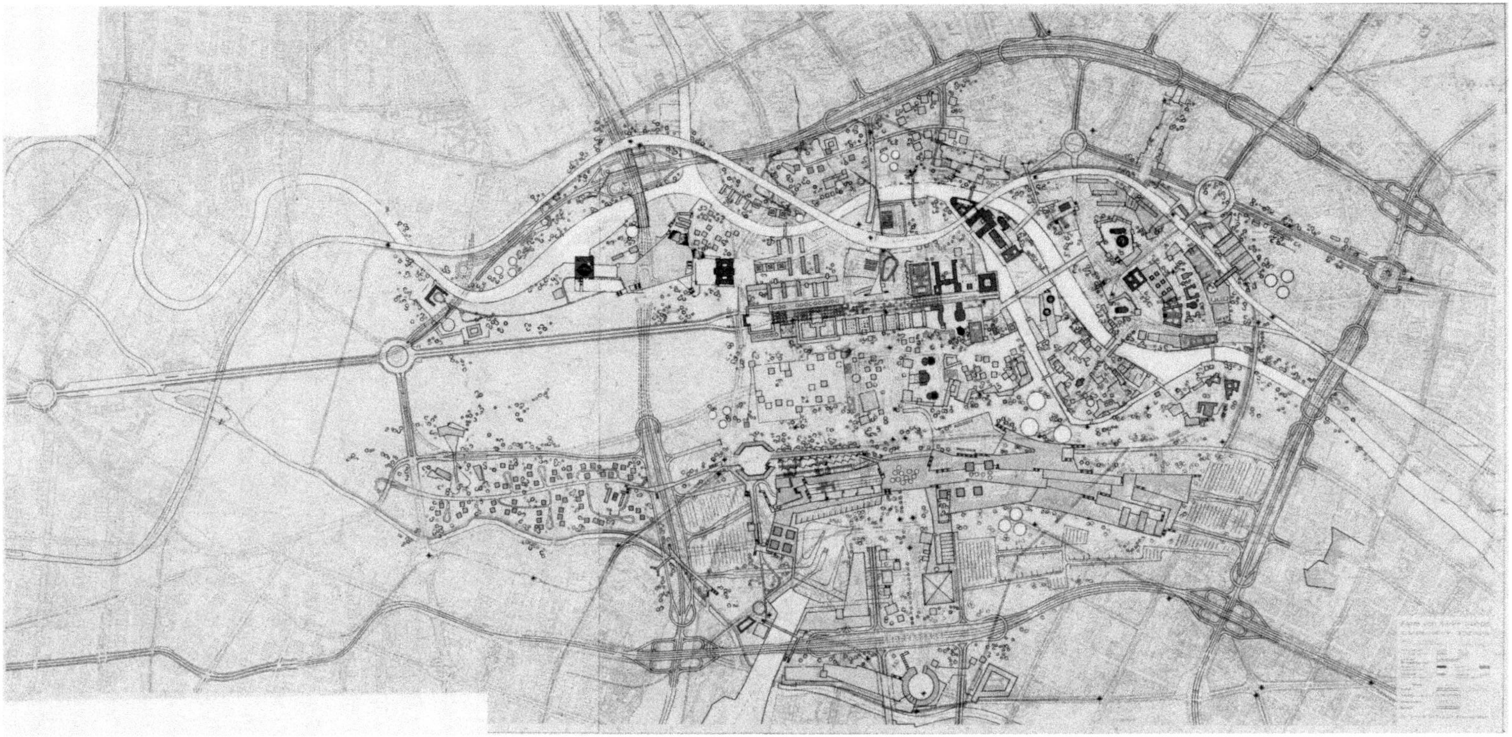

8.5 Hans Scharoun and Wils Ebert, proposal for "Hauptstadt Berlin," 1958. © 2014 Artists Rights Society (ARS), New York/VG Bild-Kunst, Bonn.

Yet these and other modernist plans put forward to develop the area were halted with the onset of National Socialist rule in Germany. Infamously, this moment produced one of the most indelible proposals for the Spreebogen site: to make over Berlin into the Welthauptstadt Germania (Figure 8.4), the seat of the Third Reich, Albert Speer proposed an immense north-south axis that would have stretched for three miles through the city. At the northern terminus, embraced by the Spreebogen, Hitler and Speer envisioned a vast domed hall that would have been the largest building in the world, spacious enough for 180,000 people and so large that Speer worried that clouds would form inside the dome. The Nazis' nightmarish visions of grandeur unified nothing but destroyed much; in their wake came a significant humbling.

Speer's plans for Berlin stood in stark contrast to the urban schemes that were proposed in the postwar years, especially the results of West Germany's Hauptstadt Berlin competition, held in 1958. During the Second World War, the Spreebogen's status as a frontier had been cemented by the fact that the boundary between the British and Soviet sectors traced its contours. As the area had been the site of tense confrontations between the Soviets and the Allied Forces from 1945 onwards, the Hauptstadt Berlin competition was compensatory, aiming to cool this particularly hot spot within the larger Cold War. The competition itself was, of course, based on a fantasy of future reunification. Hans Scharoun and Wils Ebert's plan (Figure 8.5) shared with many of the other competition proposals an insistence that only wide-scale demolition could help prepare the landscape for the new Berlin. Here, the Spreebogen area served as the boundary between the inner city and the northern suburbs, occupying the northwest corner of the urban center. Le Corbusier submitted a design that drew obviously on his earlier city plans, such as the Ville Contemporaine, the Ville Radieuse, and the Plan Voisin, in which the urban environment was divided and separated into specific functional spaces, although he tempered his vision in Berlin with the addition of lower structures and a relaxed hierarchy of forms. In fact, many of the Hauptstadt Berlin proposals drew on avant-garde plans from the 1920s, like Häring's or Poelzig's, in which high-rise slab buildings were surrounded by ample green space. Alison and Peter Smithson submitted a complex proposal in which the devastated core of the city would be remade into a multileveled, bustling city center. Hovering inside the Spreebogen would have appeared a vast, disc-like civic building that paid homage to Berlin's Expressionist history and served as a sort of beacon on the skyline, while the highways and automobiles surrounding it would provide the energetic pace of life that the Smithsons saw as central to a vibrant downtown. All of the competition entries had in common an emphatically progressive bent, annihilating the old urban fabric to make way for the new.

But like so many before them, these architects would not see their plans realized. On the morning of August 13, 1961, Soviet guards began to close the border between East and West Berlin starting just south of the Spreebogen at the Brandenburg Gate. What had previously been an invisible if nonetheless official boundary between east and west was made all too visible in the coming years, as the military of the German Democratic Republic (GDR) cut a wide swath through the heart of the city to build what they termed an "anti-fascist protective rampart." Running along the eastern half of the Spreebogen, the no-man's-land that surrounded the Wall's multiple barriers disrupted movement and charged the area with a newly intensified ideological content. From the 1960s through the 1980s, the Spreebogen became not only the actual site of the city's political division, but also the international image of the Cold War, with its interminable face-offs across boundaries. News of failed attempts to traverse the Wall and cross this frontier, many of which took place at the Spreebogen and not infrequently resulted in death, were met with fear and anger in the West as evidence of the barbarism of the GDR. Upon the fall of the Wall in 1989, jubilant Berliners crowded the Spreebogen, crossing a boundary that had left its mark in the form of the no-man's land created by the Wall.

Yet it would be several years before plans for the site would materialize and the void near the Spreebogen could be reframed. The beginning of this long planning process came in 1992, when the Bundesrepublik held a public, international competition for the development of the Spreebogen to house the government district. Officials agreed that the urban planning of Bonn, which had housed the West German government during the postwar period, could not offer a viable model for the reunified government, since its planners shunned any structure that might suggest immodest political content. Axel Schultes described Bonn as a place where buildings "lay down comfortably, like cows on a pasture" (in Zwoch, 1993, p. 49) and he echoed what many in the Bundestag believed: that the Spreebogen required a clear, self-assured statement of the new national identity. For the first time since 1945, words like "monumentality" and "order" were tentatively brought out of mothballs, replacing Bonn's incantatory insistence on "transparency," and a newly emboldened plan was solicited to express Germany's self-possession to an international audience. As Rita Süssmuth, at the time the president of the Bundestag, stated in the competition brief:

> *The German Bundestag wants to meet the demand for a transparent and efficient parliament which is close to its citizenry. It is open to the outside and conceived as a place of integration and as a centre and workshop of our democracy ... At the same time it strives to reunite the city halves, divided for decades, in a new spatial and structural order, and thus restore its social and urban identity. (Zwoch, 1993, p. 7)*

But with an international audience came international scrutiny, and a heightened sense on the part of the Bundestag that consensus for the master plan should be reached openly and democratically ("transparency" remaining a powerful concept, even if not the motivating architectural idiom). The Berlin Senate hosted a series of events they termed the "City Forum," which provided the public an opportunity to air its opinions on the new plan. At the same time, the Senate published and widely circulated a series of reports on city planning and architecture. Besides encouraging community participation in the process and fostering an atmosphere of openness around federal planning activity, these gestures, for some critics, offered the government a certain amount of protection against what they envisioned as inevitable blowback. They would merely create a plan according to the consensus of the people.

The entries submitted to the so-called "International Competition for Ideas on the Urban Design of the Spreebogen" demonstrate the variety of responses devised by international architects to manifest political unity in urban form. The brief called for a master plan of the area that would include buildings for the parliament offices, a chancellery, a press office, a press club, the German Parliamentary Association, and a federal council. Many firms, such as the Los Angeles-based Morphosis, attempted to find sincerely new ways of unifying the site. In their scheme (Figure 8.6), circulation acted as a dynamic gathering force, while government functions were contained in long, curved high-rises that echoed the course of the river. Others, like the Portuguese firm ARX, used Deconstructivist syntax to reflect the impossibility of centralized, hierarchical planning. As the architects stated, "This notion of center, related to the specificity of the program, which tries to define a new center for Germany, is also a typical romantic ideal of past centuries" (Zwoch, 1993, p. 62).

What these plans had in common was their commitment to the use of a late modernist idiom, however fractured and non-hierarchical. Yet Axel Schultes and Charlotte Frank's winning proposal cannot be so easily classified, as its critical reception attests, having been labeled variously as "hypermodern," "supermodern," "postmodern," "classical," "romantic," and "baroque." In Schultes and Frank's master plan, the Band des Bundes appears as a long, rectilinear form with curved ends. The architects suggested that each building would be constructed from concrete and glass, thus creating a provocative play between lightness and monumentality. The Band would run perpendicular to the footprint of the Reichstag building, twice crossing the Spree River with footbridges. Offices for the parliament and its committees are housed in the buildings to the east, balanced literally and figuratively by the Chancellery to the west; a public park lies to the north, and the new central train station ensures plenty of lively pedestrian traffic through the area. To the south, the Platz der Republik is defined by the Reichstag and the addition of a new building for the council of the 16 federal states. To counterpose the somewhat fortress-like appearance of the Band, the architects intended to place a "Civic Forum" for public gatherings at its center. This forum, Schultes felt, would be the most important space in the federal district; it was to contain a variety of architectural spaces, including a library, a cafe, and a sort of amphitheater in which citizens could demonstrate or simply hang out, all in the name of, as Schultes put it, "heal(ing) themselves of the malady of being German" (quoted in Wise, 1999, p. 62).

In describing their proposal to the jury, Schultes made clear his preoccupation with architecture's structural relationship to language, a notion that had been invigorated by much postmodern theory, from Charles Jencks to Peter Eisenman. For Schultes, the job of the architect is to explore not only what the built environment says, but also how it says it. In his text for the competition, Schultes played with the multiple significances of the word *Bund*: of course, it refers to the German federation, but it also indicates a link or a bond between disparate objects

8.6 Morphosis, proposal for Spreebogen, 1992. Image courtesy of Morphosis Architects.

or concepts. (It is worth noting that such semantic layering is characteristic of allegory in both its classical and postmodern guises.) He further claimed that the political body of the Bund would offer the "missing link" (Zwoch, 1993, p. 49) needed to create coherence both in politics and urbanism; in other words, the German Bund would provide the linguistic connection necessary for the Spreebogen to be a legible text. Further, Schultes concisely stated during the competition that he had no interest in attempting to communicate a single narrative with his firm's plan. In his presentation to the Bundestag, he quoted the neo-rationalist architect Aldo Rossi, saying, "In our times, our capacity for synthesis is broken; we can only offer fragments." Given this fragmentation, Schultes suggested that what was needed for the government district was not a single idea but rather what he called a "Spreebogen convention," a sort of urban lingua franca. Only within this convention, he argued, could a viewer "string the virtues of each place together" (Zwoch, 1993, p. 49).

For some critics, Schultes and Frank's design represented the triumph of modernism at a site where modernist schemes had continually failed to materialize. The slab buildings of the 1920s reappear in their proposal arranged like dominos into one monumental row, in a Weimar-inspired composition of tense, abstract geometry. Likewise, the plan's indebtedness to the 1960s has been repeatedly invoked, not only in the obvious influence of Louis Kahn and later Le Corbusier, but also in the Band's recollection of 1960s megastructures such as Superstudio's

Continuous Monument (Sonne, 2006, p. 209). The Italian group Superstudio termed their concept "an architectural model for total urbanization," and their satirical idea was that this vast gridded superstructure would eventually cover the entire surface of the globe, eliding urban difference and leaving the world pleasantly featureless. Certainly, these twentieth-century plans seem embedded in the very concept of the Band. However, others have observed that Schultes and Frank relied on many of the same planning principles—a strong emphasis on symmetry, axes and axial links, a clustering of government buildings into one more or less single-use area, and formally landscaped open spaces for gatherings and recreation—as did many planners from the nineteenth century who proposed designs for the Spreebogen, from Peter Lenne to Karl Friedrich Schinkel. To be clear, the *Band des Bundes* is not a pastiche of these earlier plans; rather, their forms resurface in the Band like debris after a shipwreck, defying easy stylistic classification.

We might begin to makes sense of this multiplicity by returning to the question of allegory. The *Band des Bundes* is not self-contained, but rather communicates by drawing a link with something unlike itself: the *allos*, or other, leading to this layered plan in which the ghosts of Spreebogens past push their way into the present. Schultes summons allegory for its open-endedness and denial of the very possibility of straightforward interpretation. This application of the term conjures Walter Benjamin, a writer whom Schultes admires and whom the architect has cited in other competition materials. Indeed, Benjamin's theory maintained a privileged status in European architectural schools in the 1960s and 1970s (the years of Schultes and Frank's training), with many students fascinated by the way in which the author had seemed to anticipate the disasters of the twentieth century (Hartoonian, 2010). In Benjamin's conception of allegorical structure, "in this guise [of allegory] history does not assume the form of the process of an eternal life so much as that of irresistible decay. Allegory thereby declares itself to be beyond beauty" (Benjamin, 1928, p. 178). According to Benjamin, it is the "common practice" of allegory "to pile up fragments ceaselessly, without any strict idea of a goal," in the very way in which Schultes' plan accrues images from the past with no resolution (p. 178). Allegory presented itself, in Benjamin's thinking, at times of profound political disquiet, and therefore proffered the most accurate literary representation of real-world experience. It seems likely that Schultes found the trauma of reunification to be just such a moment. Benjamin characterizes literary allegory in architectural terms in a way that would plainly resonate with Schultes, born in war-torn Dresden in 1943: "Allegories are, in the realm of thoughts, what ruins are in the realm of things" (p. 178).

It is telling, then, that Schultes would describe his master plan as an allegory of unification, not a high modernist symbol or metaphor. In Schultes and Frank's plan, unity, whether political, historical, or urban, is manifested as a desire rather than as reality. The band of federal buildings does not represent a mythical post-Wende German state, but rather invokes the yearning subtending the very idea of unification. The band makes its own fragility clear, showing that the connections it makes are delicate and contingent; the Bund is the "missing link" that is nonetheless temporary and vulnerable to the vagaries of history. Schultes

8.7 Spreebogen site, 2007. Deutsch: Bundeskanzleramt in Berlin from Roof of the Reichstag building. Photograph by Arnold Paul. Licensed under the Creative Commons Attribution-Share Alike 2.5 Generic license.

invokes allegory not to show the inviolability of reunification, but rather its frailty. He refers to the Spreebogen's past not to create a heroic narrative culminating in the present, but rather to show the process of breakages through which the present has been reached. Schultes' master plan shows us that, in his view, the only form possible at the Spreebogen is one that exposes the site's various ruptures instead of overlaying it with a false veneer of unity. It is only in this manner that the Spreebogen's multiple histories can be "strung together."

But I have been discussing the winning proposal of 1992, not the actual urban space that exists today (Figure 8.7). Since its completion in 2006, Schultes and Frank's design has been met with what can only be described as general disappointment; few Germans have been satisfied with it, and critics are divided between those who claim that the *Band des Bundes* is too imposing for a country with a past like Germany's and those who find it bewilderingly self-effacing. Certainly, the master plan as it was built has a curiously unfinished quality, the result of a series of budgetary cuts and politically motivated alterations to the original scheme. For example, there was strong government pressure to raise the parapet of the Chancellery above the rest of the Band, which significantly changes the architects' original concept of balance. The federal council is now housed in an early-twentieth-century mansion nearby, leaving the Platz der Republik as an ill-defined clearing that awkwardly gives way to the Tiergarten. The Chancellery and its garden, both of which Schultes also designed, cross the Spree but with a far shorter enclosing wall than originally planned, so that it is challenging to detect the garden except in an aerial view; and the Marie-Elisabeth-Lüders-Haus, the

8.8 Band des Bundes, 2007. Deutsch: Berlin—Spreebogen—Die beiden Gebäude Marie-Elisabeth-Lüders-Haus (links) and Paul-Löbe-Haus (rechts) werden durch zwei übereinander liegende Brücken ("Jakob Mierscheidt Steg" genannt) über die Spree miteinander verbunden. Die obere Brücke ist nur für Parlamentarier, die untere ist für jeden frei begehbar. Photograph by Andreas Steinhoff.

building that forms the eastern-most section of the Band, stops abruptly short of where Schultes and Frank originally intended. Perhaps most importantly, the Civic Forum has all but disappeared as an architectural space and is now present only as an uncertain caesura in the Band, with landscaped *allees* of trees, fountains and vehicle traffic running constantly through it (Figure 8.8).

When plans for these changes were finalized in 2001, Schultes defended his firm's original concept against a barrage of bad press, asserting that his feelings during the Spreebogen project hovered between "frustration and elation," more frequently becoming frustration (quoted in Haubrich, 2001). One journalist for *Die Welt* spoke with Schultes about his unrealized dreams for the project and reported that Schultes could only "appeal to the imagination of the audience to imagine this band at its best, and complain again that the 'heart' of his 'political-architectural concept,' the 'forum' between the Chancellery and the Bundestag blocks, has now been deleted silently." Therefore, concluded the author, "here, only the void will rule" (Haubrich, 2001).

But this, too, is telling. In essence, Schultes got the allegory he never wanted, one that narrates, as Craig Owens described the postmodern "allegorical impulse," "its own contingency, insufficiency, lack of transcendence" (1980, p. 80). As Owens noted, such an impulse "tells of a desire that must be perpetually deferred" (p. 80). Schultes had imagined his design as a legible—at least, legibly allegorical—text;

but in the built *Band des Bundes*, Schultes and Frank's proposal becomes one more fragment of the past palimpsestically pushing its way into the present reality. The connections between architecture and literature strengthen through this process of layering; Owens suggests, "In allegorical structure, then, one text is *read through* another, however fragmentary, intermittent, or chaotic their relationship may be; the paradigm for the allegorical work is thus the palimpsest" (p. 80).

Ultimately, then perhaps Schultes and Frank's master plan is allegorical in way in which they never intended, that is, in the sense that all master plans are exercises that have more to do with literature, and with a sort of literary figuration, than they do with architecture. They are panoramic views that require the viewer to "string together" historical moments in order retroactively to invest the fragmented site with specific meaning, as the viewer is required to do with Schultes and Frank's Band des Bundes. In Benjamin's terms, master plans function as "the expression of an Idea" (p. 161), rather than as lived experience. The master plan inherently exists beyond the view from the ground, perceptible only from a vantage point outside itself (from above, from the page). In theory, the pedestrian will experience or intuit the logic of a master plan without access to this totalizing view, but in fact, the pedestrian is always blind to the totality of the master plan and its unbuilt pasts. Once within it, the viewer cannot see it. The plan marks the boundary between architectural fact and architectural fantasy.

In fact, that the city of Berlin itself—and even more intensely, certain historically burdened zones within it, such as the Spreebogen—are troped as boundaries is by now a truism bordering on the cliché, thanks to the twentieth century's most notorious zone marker. Since the middle of the twentieth century, novels, films and works of art have thematized Berlin as the home of the so-called "inhuman frontier," and fascination with the divided city continues unabated even as the *Wende* retreats further in time. The Wall's various anniversaries continue to attract international attention; in 2009, events were held to mark the 20 years that had passed since the fall of the Wall, in 2011, international media channels observed the 50 years that had elapsed since its construction, and so on. As well as the seductive visual and literary backdrop the wall proffers, the problems created by its demise continue to be discussed. For example, a symposium held at the Yale University School of architecture in 2013, titled "Achtung: Berlin," examined the urban impact of the traumas of the last century, while a temporary installation called "Blurring Boundary" recently appeared at the Wall Memorial as an attempt to overcome the still-tangible presence of the Wall within the city. In all cases, Berlin is presented as an unresolved territory, a frontier whose terms of occupation remain unclear, and a site of tension and confrontation across borders. Schultes and Frank's Spreebogen master plan, then, functions allegorically to make meaning from this borderland, not as a contained and unified symbol, but as a fragmented collection of discrete objects. Therefore, the *Band des Bundes* in particular and the master plan in general might be seen as a purely literary device, a formal experiment that creates a narrative by linking unlike things. Yet such an assembly of disparate spaces is guaranteed to point primarily to its own implacable decay, as the links weaken over time. At the Spreebogen, the border remains a void, not the place where governments, militaries, histories, or narratives cohere, but rather the zone where those fragments pile up, where they fail to become integrated, gazing blankly at each other across the frontier.

REFERENCES

Barthes, R. 1964. "The Eiffel Tower." In S. Sontag (ed.) 1982. *A Barthes Reader*. New York: Hill and Wang, pp. 236–50.

Benjamin, W. 1928. *The Origin of German Tragic Drama*. Translated by John Osborne, 1977. London: NLB.

Hartoonian, G. 2010. *Walter Benjamin and Architecture*. London: Routledge.

Haubrich, R. 2001. "Hier regiert die Leere." *Die Welt*, January 31, 2001 [online]. Available at: http://www.welt.de/print-welt/article431126/Hier-regiert-die-Leere.html (accessed 10 February 2012).

Koshar, R. 1993. *From Monuments to Traces: Artifacts of German Memory 1870–1990*. Berkeley: University of California Press.

Owens, C. 1980. The allegorical impulse: Toward a theory of postmodernism. *October*, 12 (Spring 1980), pp. 67–86.

Sonne, W. 2006. "Berlin: Capital under Changing Political Systems." In D. Gordon (ed.) 2006. *Planning Twentieth Century Capital Cities*. London: Routledge.

Urban, F. 2004. "Recovering essence through demolition: The 'organic' city in postwar West Berlin." *Journal of the Society of Architectural Historians* 63 (September), pp. 354–69.

Wise, M. 1999. *Capital Dilemma: Germany's Search for a New Architecture of Democracy*. New York: Princeton Architectural Press.

Zwoch, F. (ed.) 1993. *Capital Berlin: Parliament District at the Spreebogen: International Competition for Urban Design Ideas*. Berlin: Birkhäuser.

PART III
Spatial Ambiguity and (Dis)Embodied Memory

9

Gorizia and Nova Gorica: One Town in Two European Countries

Tina Potočnik

"In different forms and ways, boundaries and borders both have to do with the modification of our real landscape, transforming the territory that we physically occupy and inhabit" (Zanini, 1997). Having Piero Zanini's statement in mind, it is worth considering how frontiers and their alterations have influenced the formation and modification of our built environment at various levels, including the urban level. Many settlements have been denoted as *border cities* or *frontier towns*, ranging from Berlin to Jerusalem, with this denomination arising from diverse historical or current political situations; yet always the border has an important impact on the structure and functioning of the settlement. The focus of this chapter is on the Slovenian-Italian border and two small towns closely connected with it, the medieval town of Gorizia (Gorica), today the center of the Italian province of Gorizia in the Region Friuli-Venezia Giulia, and the Slovenian town of Nova Gorica (*New* Gorica), erected after World War II and now the center of the Primorska region. Their parallel lives have been the subject of many studies, from historical, urban, architectural, socio-political, economic and geographical perspectives (Vuga, 1990; Jurca, 1990, 1997; Torkar, 1990, 1999; Angelillo, Angelillo and Menato, 1994; Zoltan, 1997 Spagna, 2003; Angelillo, 2004; Marin, 2007; Cattunar, 2007; Di Battista, 2011). Yet increased interest in this area, especially in the context of the integration of the two towns with the abolition of borders in the process of the enlargement of the European Union and the Schengen Area, raised some new questions related to the changing characteristics of the (former) frontier.

With this chapter I attempt to answer some of these questions, specifically how the new boundary, drawn after World War II and whose shifting characteristics and perception are conditional on the changing historical circumstances, influenced the formation or transformation and the development of these two urban settlements in the frontier zone. Another issue examined in this chapter is how the latter were affected by the change in the character of the border after the fall and dissolution of Yugoslavia, after Slovenia gained its independence in 1991, and also following the abolition of boundaries and the process of enlargement of the EU. I discuss the impact of the border on the (trans)formation of Gorizia and Nova Gorica by correlating the characteristics of the border, especially its degree of *openness* and including its socio-political and ideological context, with the attitudes arising from the most important urban development plans on both sides of the border, as well as with the main tendencies for the development of the conurbation in the last decades. Furthermore, I shed light on the creation of a specific identity of built structures in this frontier zone due to the frontier as it was set after World War II.

THE *OLD* GORIZIA AND THE NEW FRONTIER(S)

The border area discussed in this text used to be the location of one settlement only, called Gorica (today's Italian Gorizia). Already in the oldest written still-surviving source mentioning this settlement that originates from 1001, there is the statement regarding the boundary, namely the boundary between two dynastic territories, i.e. the emerging county of Gorizia and the patriarchy of Aquileia. At the same time this was a transitional area between the Balkans, Central and Western Europe and the Apennine Peninsula; thus, due to its geographical position in the eleventh and twelfth centuries, Gorizia became an urban settlement of greater importance for the wider region. This medieval town under the castle obtained market rights in 1210 (Gorica, 1989, p. 279) and was under the rule of the Counts of Gorizia who, "sandwiched between the burgeoning Habsburg Empire to the north and west, and the expansionist Venetian Republic to the south / … / ruled a patchwork of territories that stretched from the eastern Tirol, across present-day southern Austria and northeast Italy and down into Slovenia and Croatia" (Morris, 2000). Yet in 1500, after the extinction of this local nobility, Gorizia was, along with their dynastic territories, incorporated into the Habsburg Monarchy (Gorica, 1989 p. 280). Consequently Gorizia became a border town.

The next important shift in the location of the border followed after World War I, but Gorizia had continued to develop, especially after 1860 when the Vienna-Trieste-Gorizia railway, i.e. the Ferrovia Meridionale or Südbahn, was built. The town came to be not only the administrative and religious but also a prosperous commercial, industrial and tourist center of the region (Codellia and Graziati, 2000, pp. 13–14; Marin, 2007, pp. 19–23). With a harmonious blend of medieval and baroque architecture, as for example in Piazza della Vittoria, the traditional center of the town originating in the Middle Ages, Gorizia began to expand and modify its appearance considerably after 1906, when the Bohinj railway, i.e. the Ferrovia Transalpina or Wocheinerbahn, was opened.[1] Further development was interrupted by World War I, which left the town largely destroyed (the Battles of the

[1] The so called North railway station was (similar to the "South railway station" or the Ferrovia Meridionale) built about a kilometer away from the medieval center, which again caused the urbanization of the space in between. With the expansion of the city center the need of the plan for urban development arose (Codellia and Graziati, 2000, p. 14; Marin, 2007, pp. 32–3).

Isonzo between the Austro-Hungarian and Italian armies that were fought in close proximity) and led to the Treaty of Rapallo in 1920. At this point the location of the frontier was changed when the treaty, which determined the border between the Kingdom of Italy and the Kingdom of Serbs, Croats and Slovenes (later Yugoslavia), assigned Gorizia together with its hinterland to Italy and relocated the border towards the east (Gorica, 1989, p. 280). The frontier zone shifted again after World War II, when the peace treaty signed in Paris on February 10, 1947 assigned the urban center of Gorizia to Italy, while its eastern suburbs with the north railway station and rural hinterland were left to Yugoslavia (Gabrič, 2005, pp. 831–2; Troha, 2005, p. 909).

The Yugoslav government, at that time restoring the damage from the war, *building* socialism and *creating* the socialist society with great élan, reacted to this situation; at first considering turning one of the existing historic settlements into the new center of Primorska, i.e. Solkan, Šempeter or Ajdovščina (Torkar, 1997, p. 15), and later considering connecting two of those settlements, more precisely Solkan and Šempeter, with a road that would run right next to the railway (Ravnikar, 1984, p. 43). Finally, the decision was made to erect a completely new town in close proximity to the lost town of Gorizia and the new border (Vrišer, 1959, p. 45; author of the article quotes oral sources; Nanut, 1972, pp. 7–8; Plahuta, 1979, p. 148).

CONTEXT OF THE BORDER IN 1947 AND PLANS FOR NOVA GORICA

Although this wider area had been a border territory for centuries, the new frontier, set just next to Gorizia in 1947, was in terms of a micro-location a completely new experience. At first local people did not take the border seriously, they did not perceive it as something that they were not allowed to cross, because there were so many everyday things, habits and needs uniting both sides; after all, Gorizia was the administrative, economic and cultural center of the agricultural outskirts at the time it was separated from Yugoslavia.[2] This connection appeared clearly in the first plans for a new town on the Yugoslav side. In the preliminary draft of the urban planning design for Nova Gorica, produced by the local architect Božidar Gvardjančič in April of 1947,[3] the idea of building an addition to the old Gorizia, an

---

[2] More on the perception of the here discussed border see Bufon (1995), Malnič (2008), Širok (2010) and documentary films *Ordinacija spomina. Corsia dei ricordi*, 2009 (Production: Kinoatelje, directed by Anja Medved), *Pogledi skozi železno zaveso. Sguardi attraverso la cortina di ferro*, 2010 (Production: Kinoatelje, directed by Anja Medved), *Moja meja. Il mio confine*, 2005 (Production: Kinoatelje, directed by Anja Medved and Nadja Velušček).

[3] The Regional Archives Nova Gorica, PANG 61, Zbirka gradiva o gradnji Nove Gorice, tehnična enota 2, Božidar Gvardjančič: The plan for Nova Gorica (April 1947). The basic is the grid street plan; streets run at right angles to the main road (the north-south axis), while secondary streets are diagonal. West of the main axis there are two rows of apartment buildings (blocks); behind them is the recreational area. East of the main axis there are public buildings (schools, administration buildings, culture), further

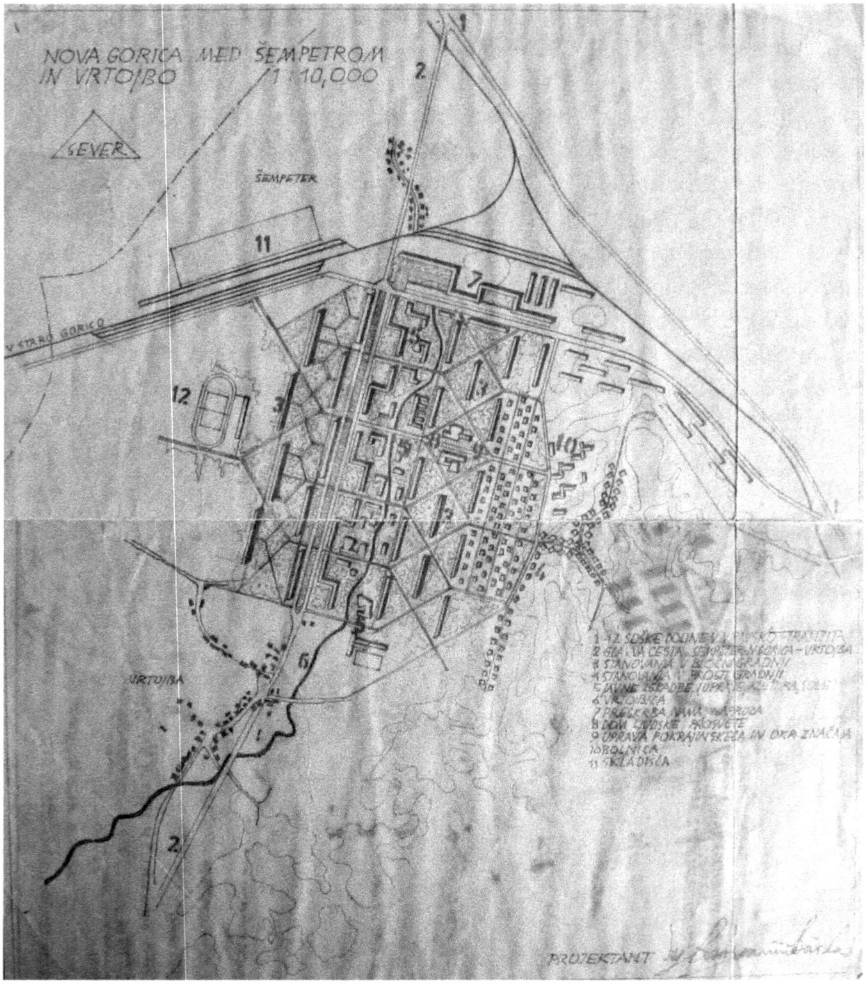

9.1 Božidar Gvardjančič: The plan for Nova Gorica (April 1947).
Reproduced by permission of The Regional Archives Nova Gorica.

extension across the border, was argued. Gvardjančič planned a new town located next to the border as a kind of neighbourhood of the old Gorizia (Torkar, 1990, p. 11). Besides the choice of the location itself, also the fact that the road connections with Gorizia would be kept and that there was also no new industrial area planned were decisions that stemmed from the Yugoslavian side's hope to regain Gorizia.

As a result of some disagreements between the local authorities, in July of the same year Gvardjančič prepared a new draft of the urban planning design for Nova Gorica.[4] Although this plan foresaw Nova Gorica in a different location, this new

away there are three rows of apartment blocks and different types of housing (smaller apartment buildings) on the outskirts of the town.

4 The Regional Archives Nova Gorica, PANG 61, Zbirka gradiva o gradnji Nove Gorice, tehnična enota 2, Božidar Gvardjančič: The plan for Nova Gorica (July 1947).

9.2 Božidar Gvardjančič: The plan for Nova Gorica (July 1947).
Reproduced by permission of The Regional Archives Nova Gorica.

town too was to be located right next to the border. The settlement was designed as a natural extension of the old town and as the latest part of the historical Gorizia, which would together form so-called Velika Gorica (*Great* Gorica). Also the main

---

Again the basic is the grid street plan. The main axis is the road that connects the old town of Gorizia and the square with some prominent public buildings. Along this axis there are all important activities and functions. North of the main axis is the residential area, with diagonally placed apartment blocks. Between the main regional road that leads along the border and the railroad is the industrial area that was not foreseen in the first plan.

axis was directed towards the old Gorizia. This reflects the hope for regaining Gorizia that was still present at that time (Torkar, 1990 p. 12).

But the characteristics of the border considerably changed in the very same year as on September 15, 1947 the provisions of the Paris Peace Treaty came into force. The border was completely closed. The wire fence was set and the frontier zone became a highly guarded area in which nothing higher than potatoes was allowed to grow. The border created between the communist and anticommunist worlds[5] split fields, cut off roads and divided families and friends; Gorizia remained *without its soul* (Anon., 1947). At this point, Minister of Construction of the People's Republic of Slovenia, Ivan Maček Matija,[6] assigned three architects, Gvardjančič, Marko Zupančič[7] and Edvard Ravnikar, the task to each produce his own draft of the urban design for the new city. The plan of Gvardjančič resembled his first two plans. He located the new town just by the border and he designed a monumental square at the crossroads, from which again one road led to the old Gorizia.[8] In the plan of Zupančič, Nova Gorica was also located in close proximity to the border. The old Gorizia was seen as the old core to which new neighbourhoods were attached—Nova Gorica on the north, Vrtojba in the middle and Šempeter on the south.[9]

The minister allegedly reproached Gvardjančič and Zupančič for linking the new town to the old Gorizia and ignoring the required 500-meter-wide protection zone along the border. As a result of its accord with the then-legal regulations and its consideration of the minister's statutory provisions, among which roughly a hundred meters of space was to be left empty along the border, Ravnikar's plan was selected.[10] If at first there was "the predominant idea that the city, dismembered

---

5 Kaja Širok (2010, p. 340), who focuses on the formation of the collective memory in the frontier zone of the Goriška region, objects to the comparison between the Italian-Slovenian border and the Berlin wall. She states that naming the here discussed border the Iron Curtain, as it appeared later (in 2004 and 2007) in media, is the discursive invention of the tradition (after Eric Hobsbawm).

6 After September 15, 1947 and the annexation of the Primorska the Ministry of Building and Urban Development and the minister Ivan Maček Matija took over the responsibility for Nova Gorica (Torkar, 1990, p. 12). The minister played an important role in determining the location and in deciding on the design. Yet, the document with the decision to build Nova Gorica was not preserved.

7 The two drafts for urban design plans were attributed to Gvardjančič and Zupančič by Alenka Di Battista (2011, p. 322).

8 There are some differences though, for example, there is a road crossing the main north-south axis, but there is no regional road along the border—this function is given to the main road through the town that connects Solkan and Šempeter.

9 The main axis that goes from north to south is a wide transit road with two tunnels. There are also two secondary roads that run at right angle to the main road, connecting the railway station on the west and leading towards the east; and there is a ring road (beltway).

10 Beside the consideration of the 500-metre-wide border strip also the relocation of the railway line (from Solkan under St Cathereine's hill) and the connection between the Soča and Vipava Valleys with two tunnels were among those provisions. The Regional Archives Nova Gorica, Fond 61: Zbirka gradiva o gradnji Nove Gorice, Pismo

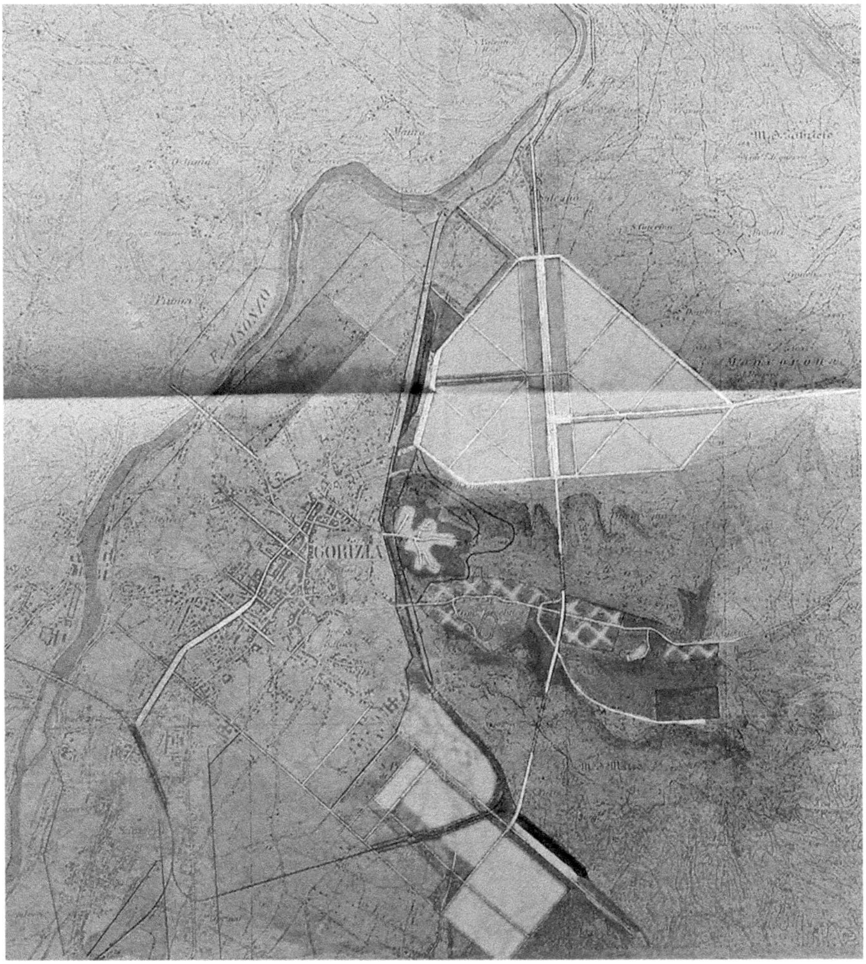

9.3 Marko Zupančič: The conceptual design for Nova Gorica (September 1947). © Marko Zupančič. Archive of the National Museum of Contemporary History (Slovenia).

by the railway and the border, should one day, or as soon as possible, be united again" later, as Ravnikar continues "(b)order security radically reduced our space along the railway with the decision to maintain a 500 m belt completely empty. Moreover, any construction had to maintain a distance of at least 80 m from this belt. As a consequence, the idea of the two eastern suburbs, which were ready to assume their role in the anticipated Velika Gorica, had to be abolished" (Ravnikar, 1984, p. 46). Ravnikar's statement indicates that the idea of a Greate Gorica had to be decisively abandoned.

B Gvardjančiča Cirilu Zupancu, Ljubljana, 4. 12. 1967, t. e. 2. It is possible that Ravnikar's political activity in the Liberation Front of the Slovene Nation also influenced the decision of the minister (cf. Kopač, 1995, pp. 212–15).

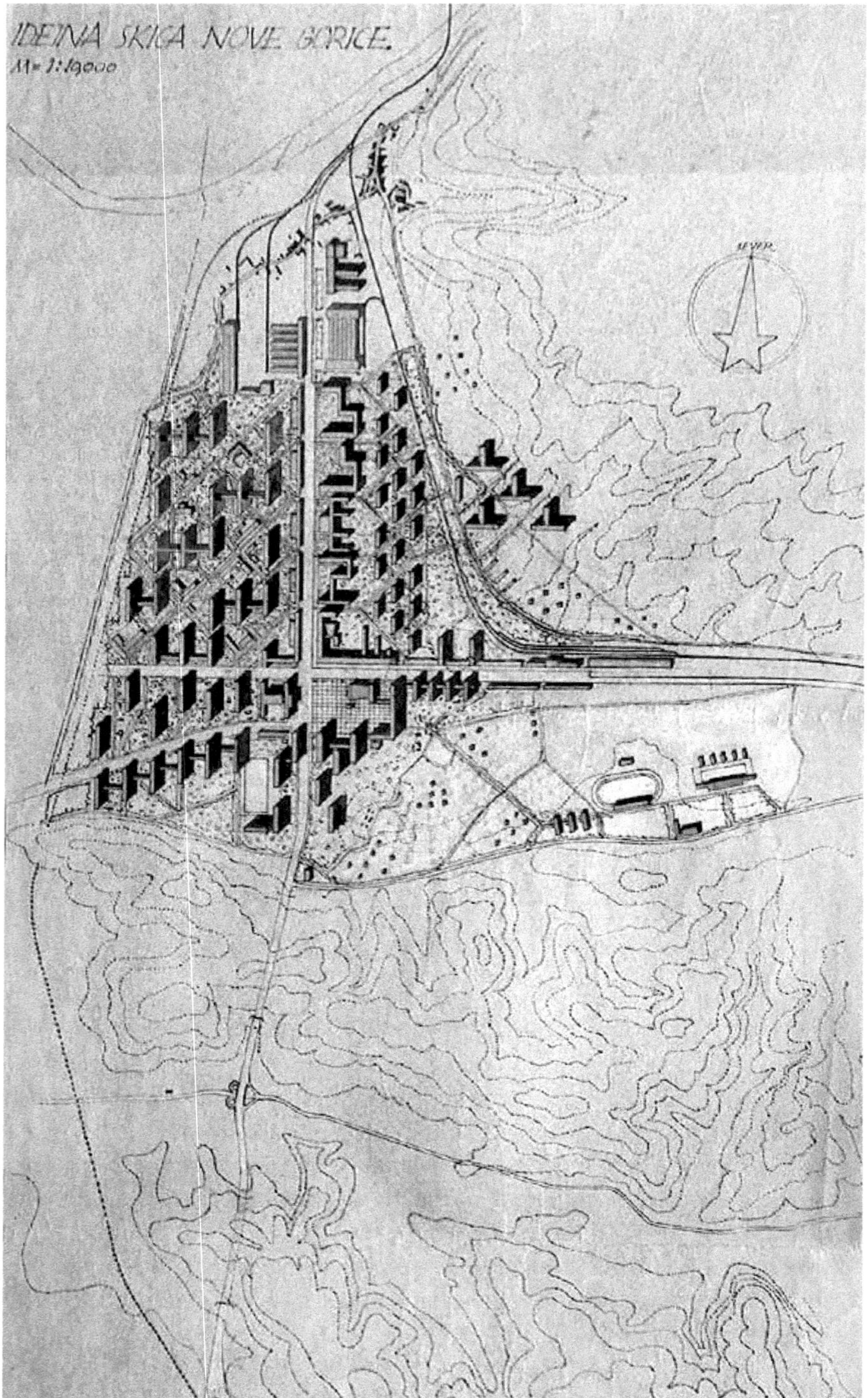

9.4 Božidar Gvardjančič: The conceptual design for Nova Gorica (September 1947).

Edvard Ravnikar, the student of well-known European architect Jože Plečnik, was the central figure of Slovenian post-World War II modern architecture and a pioneer of modern urbanism in Slovenia.[11] In 1939 Ravnikar worked in the studio of Le Corbusier; in fact, Le Corbusier's and CIAM's urban doctrine was a model for his plans for Nova Gorica, dating from the years 1947 to 1950.[12] His urban design plan for the town, prefigured along the north-south arterial road, is similar to the plans of Gvardjančič based on the idea of a garden city, with the distribution of basic functions, wide roads that cross at right angles and vast platforms of apartment blocks and modernist palaces surrounded by greenery, with an abundance of air and light. Ravnikar's new town, built as a demonstration of the superiority of socialist society, was not only to replace Gorizia, but also to symbolically overshadow it. "When as a result of peace negotiations Gorica was definitely lost and dreams of a Yugoslav Gorica went up in smoke, / … / (w)e were told to build something great, beautiful and proud, something that would shine over the border" (Ravnikar, 1984, p. 46). It is interesting, however, that there is still a connection with Gorizia in Ravnikar's plan, namely the linkage between the organizational network of the northern railway line on the eastern outskirts of Gorizia and, in particular, the diagonal road beginning in the central area of the town, more precisely at the square in front of the Town Hall, crossing some hundred meters of empty frontier zone and ending in a market near the main square under the castle of Gorizia.

While on the east side of the border a modern town foreseen as a built materialization of the new society was being constructed (although Ravnikar's plan was only partially realized[13]), Gorizia, which remained without the agrarian hinterland on the east side of the border, was experiencing a decline in economic, cultural, demographic, and other fields.[14] The municipal government stayed true to the tradition in urban planning in the sense that it retained the plan of the architect

---

11 For more on Edvard Ravnikar and his work see Krečič, Mušič and Zupan (1995), Koselj (1995), Ivanšek (1995), Žnidaršič (2004), Bernik (2004) and Vodopivec and Žnidaršič (2010).

12 Ravnikar began dealing with Nova Gorica in 1947 with his students France Ivanšek and Franc Šmid. The urban design plan from 1947, on which the construction of Nova Gorica began, was published in 1948 (Mušič and Ivanšek, 1948). In 1950 the general plan for the urban development was made. After 1950 Ravnikar took no further part in the development of Nova Gorica (Torkar, 1990, p. 16).

13 Construction followed partial plans and two subsequent urban development plans from the 1950's. Those were the plans of Gvardjančič and of Viljem Strmecki, which directed the development in the opposite direction of that as foreseen by Ravnikar. By Ravnikar's plan the main road, six apartment blocks and the building of the local authority were constructed (Plahuta, 1979, p. 156).

14 Because the connections with the eastern and northern countries which were still very strong under the Habsburg rule were cut, the era of gradual stagnation of the town began. The economic, commercial balance that Gorizia scarcely regained after World War I was lost. Italy tried to improve this situation with a law that foresaw a duty-free zone in this area but in the long-run it did not succeed (Spagna, 1973, p. 15).

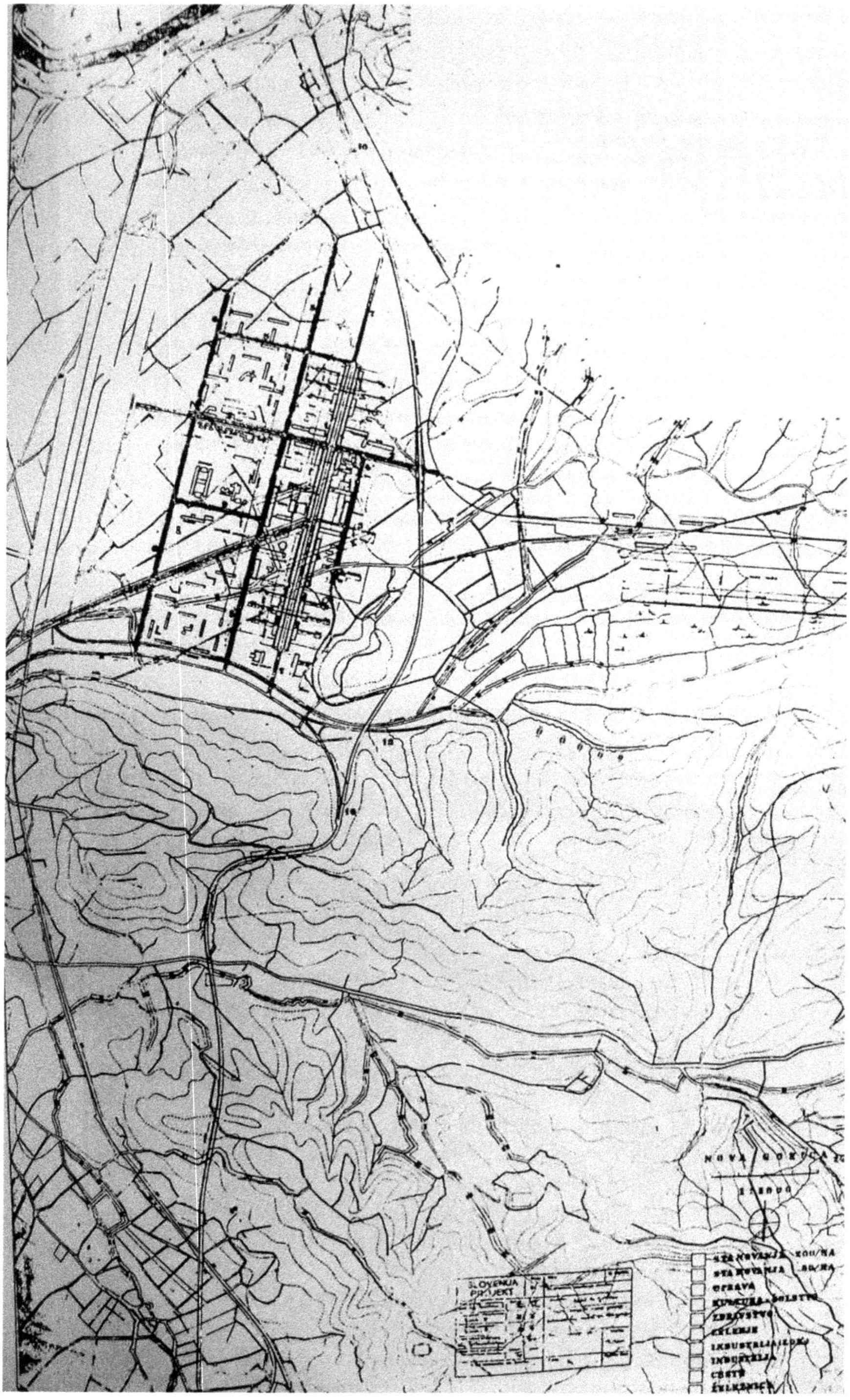

9.5 Edvard Ravnikar (in collaboration with Franc Šmid): The conceptual design for Nova Gorica (1947). Mušič, M. and Ivanšek, F. (1948).

9.6 Photograph of the 3D model of Nova Gorica (1948).
Reproduced courtesy of Muzej Za Arhitekturo in Oblikovanje.

Maks Fabiani (1865–1962) from 1921[15] (with some additions) for the next 40 years. Fabiani prepared an urban development plan for Gorizia after World War II, in which he focused above all on the renovation of the castle, its surroundings, and Piazza della Vittoria. In his plan the road connection and the railway disappear where they reach the border and the new town on the other side is ignored (Fabiani, 1949a, p. 2, 1949b, p. 2, 1949c, p. 2). Only in 1962 was an Italian architect, Luigi Piccinato, given the task of producing a new general plan for the urban development of the city. What is important for the relationship between the frontier and the (trans) formation and development of urban settlements is that Piccinato's plan from the year 1966[16] foresaw the future development of Gorizia in international traffic and trade, yet not in the direction west-east. Thus the plan basically tried to direct the growth of the town towards the south-west and south-east. This isolation from Nova Gorizia was emphasized with the planned highway from Venice through Gorizia and Klagenfurt to Vienna, i.e. in the south-north direction.

Though it was technically advanced, the plan failed to prevent the increasing isolation of the old Gorizia and the escalation of its regression (Marin, 2007, pp. 61–2;

[15] The well-known Slovenian architect Maks Fabiani prepared the first urban development plan for Gorizia, which was implemented after World War I in 1922 (Marin, 2007, pp. 41–5; Codellia and Graziati, 2000, p. 36).

[16] The general urban plan was made due to the new law that originated in 1942 and came into force for Gorizia in 1954, which required a general regulation plan of the town (Urbanistica e edilizia, 1986, pp. 967–8).

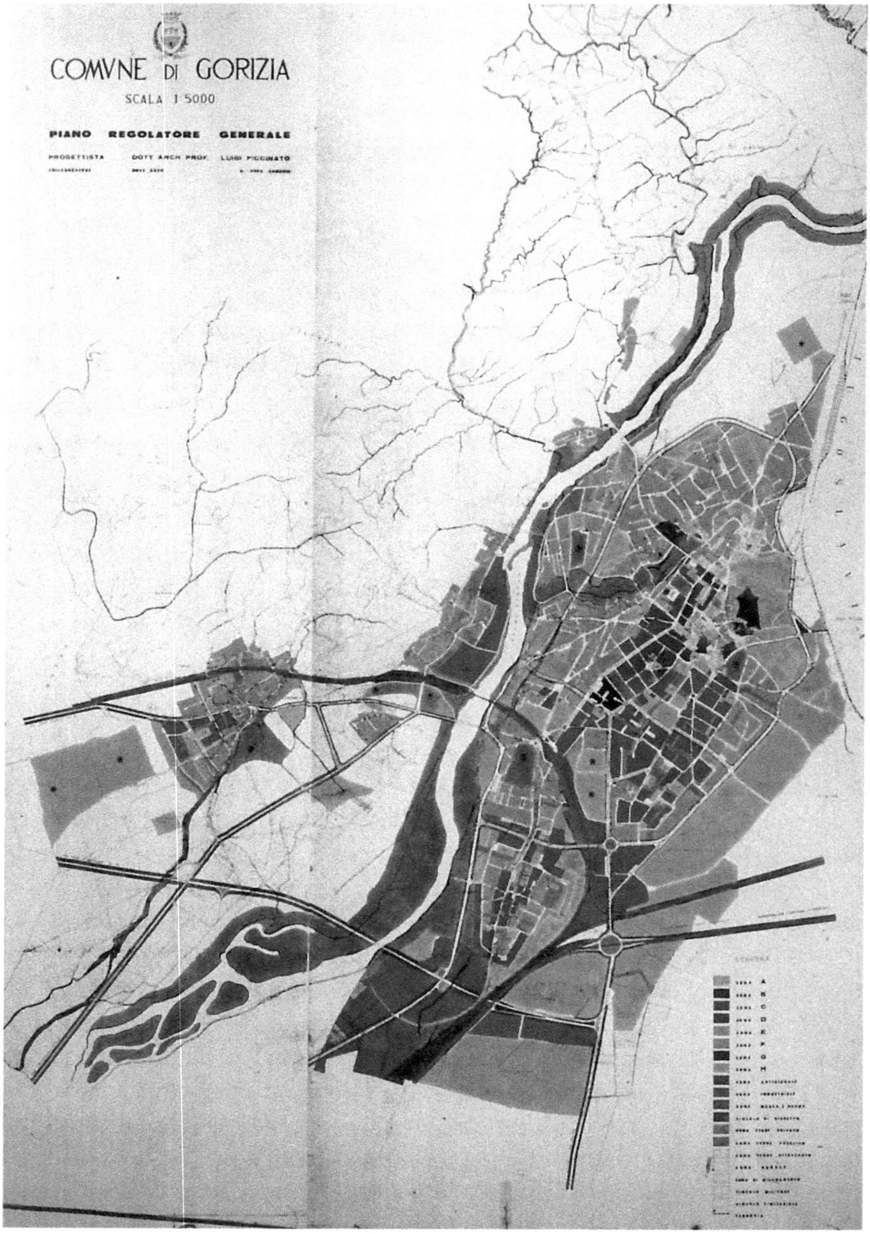

9.7 Luigi Piccinato: The general plan for urban development for Gorizia (1966). Cattunar, A. (2007).

Spagna, 1973, pp. 18–20). At the same time the development of Nova Gorica was hindered, also due to the *impassability* of the border. Consequently, Nova Gorica did not reach its heyday until the 1980's, when the gambling industry was set up in town (Nova Gorica, 1994, p. 12). Hence the border caused the erection of a new town, the regression of the old one, and stymied development of both settlements in the frontier zone, all this in spite of the modern urban planning on both sides of the border, which, as Ravnikar states, became:

> *a weapon in the national and political struggles. The momentum that existed then was full of the pathos of great historic changes in urban planning and architecture. They were seen as alternative to the war and armament, as something that would resolve all essential problems of human society / … /, help abolish the differences arising from borders, races, and classes / … /. (Ravnikar, 1984, p. 46)*

THE FORMATION OF THE CONURBATION

In the mid 1950s, i.e. already before Piccinato's plan, which ignored the existence of the modern town on the other side of the border and developed in a different way as proposed by Ravnikar, the border gradually became more open as several agreements between Yugoslavia and Italy set the basis for the new neighborly relations.[17] Two decades later, with the Treaty of Osimo, signed in 1975,[18] the Yugoslav-Italian border seems from today's perspective to have become the most passable (open) border of that time among the existing socialist and capitalist countries in Europe. The cooperation between Gorizia and Nova Gorica significantly expanded after this year. Joint commissions for cooperation in various fields were set up (economy, culture, education, sports, environment, urbanism etc.), among which the Town Planning and Environmental Protection Commission was the most active one, coordinating urban planning on both sides of the border (Vuga, 1990, pp. 59–62). In this context, the idea was accepted to connect two streets, Erjavčeva Street and San Gabriele Street, turning them into a promenade along the shortest route between the two city centers (Šušmelj, 1997, pp. 17–18). Opening the border crossing for pedestrians on this promenade was a very important step in the elimination of the sense of the frontier zone. In addition, other border adjustments were made that rectified inconsistencies of the border set in 1947, following the logic of regulation and the needs of daily life (Vuga, 1990, pp. 59–62).

Finally, after the fall of the socialist regime and the attainment of the independence of Slovenia in 1991, the border no longer separated two ideologies, but only two countries. Again, new possibilities for the convergence of the two cities emerged in various sectors, such as culture, economy and infrastructure. In 1991 a proposal for the convergence of the two towns was signed, offering many concrete initiatives, including joint urban planning, that would facilitate their cooperation (Šušmelj, 1997 p. 23). In the framework of these tendencies numerous

---

[17] The border was completely closed until 1949, when a treaty was signed in Videm (Udine). In the middle of 1950s another two treaties were signed—one in Videm and another in Gorizia, which dealt with the local border traffic. The relations between Italy and Yugoslavia improved and actuated the flow of goods, passengers and ideas (Gorica, 1989, p. 280).

[18] This treaty, signed on 10th November 1975, set the final course of the border that was not exactly determined in 1947.

studies and plans for urban development were carried out that offered various solutions to regulate the frontier zone.[19] The plans, of which only a few were realized, suggested integration at the urban level. Their aim was to act beyond the frontier, to obliterate it from the urban structure. The new urban development plan for Gorizia as created by Italian architect Augusto Cagnardi that came into force in 2001 and foresaw the reurbanization of the frontier zone followed this model, as did the winning plan for the development of the frontier zone by Slovenian architect Vojteh Ravnikar that foresaw along the border a new quarter of Nova Gorica, which however was never realized (Spagna, 2003, pp. 13–15).[20] It seems that the crucial moment in these efforts occurred when Slovenia joined the European Union in 2004, as this occasion was honored by both the Italian and the Slovenian municipal administrations' decisions to build a new joint square in front of the railway station in Nova Gorica, where the two towns had before been divided by the border line. On the new Square of Europe, which extends over both sides, in Slovenia as well as in Italy, the former border is only symbolically indicated by a line on the ground, which transversely intersects this open space. This square became the place of the closest contact between both towns.

When Slovenia joined the Schengen area in 2007, the free movement of capital and people between Gorizia and Nova Gorica was launched. The border was abolished; Gorizia and Nova Gorica, which in fact had grown together before that, became a *transnational conurbation*. However, the memory of the frontier remains in the urban structure, above all in the progressively reduced density and in the presence of the less important and attractive activities that are typical of suburban areas. These are indeed located on the outskirts of Nova Gorica but at the same time in proximity to the old center of Gorizia. The main reminder of the frontier in both urban structures is therefore the fact of the existence of the conurbation itself and the structure of this conurbation, especially around the former border line where the margins of both towns have become the center of their conurbation.

---

[19] In 1991 a group of Slovenian architects AxA (Dare Poženel, Vinko Torkar and Niko Jurca) prepared a document regarding the joint development of the two towns into a "Great Gorica" (Jurca, 1997). An Italian group of architects (Alfonso and Antonio Angelillo, Chiara Menato) introduced their own idea of urban development of the frontier zone on the EUROPAN 2 competition in 1991 (*Progettare il confine per un territorio senza barriere*, 1991). Furthermore, the themes related to the frontier zone between Gorizia and Nova Gorica have also been included in the European programs of INTERREG (Marin, 2007, pp. 97–9). In 1996 the first bilateral document on joint development was signed by the municipal administrations (*Progetto di riconciliazione tra abitanti di Gorizia e di Nova Gorica*, 1998). Additionally, in 1997 also the international symposium with the focus on the planning in the frontier zone took place in Gorizia (Pozzetto, 2004, pp. 36–7).

[20] See also the competition entry 66899 by architect Vojteh Ravnikar: *Urbanistična ureditev območja ob železniški postaji v Gorici*, 2003, p. 4.

CONCLUSION

The new frontier, mapped out in 1947, led to the coexistence of the two paradigmatically different towns; on one side a rationalist town, erected due to its instantaneous historical circumstances, a town that came into existence because of the frontier and was formed by it and, on the other side, a historical town, a town which arose over almost a millennium of history and was, in contrast, conditioned by the space of flows and not the limits that later pushed it into regression.

The impact of the border, which was initially not seen as something impassable, became, as it was completely closed for some time, highly disjunctive also in terms of the urban development of the two settlements in the frontier zone. Through the study of unrealized and realized plans for Nova Gorica and those that transformed the *old* Gorizia, the impact of political interests (not only in the socialist state) and especially geopolitics on town planning and thus on the form and image of the two settlements reveals itself in another light. This (political) frontier, and its ever-changing (political) character, directed urban development by justifying the essential role of politics in town planning; this meant that culture, tradition and design issues were kept in the background. A complete antipode to the first, never-realized plans for Nova Gorica by Gvardjančič and Zupančič, that still strongly relate to the old Gorizia, was Ravnikar's plan, selected as the most appropriate by the Minister of Construction, which (due to the changed historical circumstances and consequentially shifted perception of the border) foresaw the new town as a separate entity. At the time of the creation of the *new town* the demands of Yugoslav politics and the perception of border as a fixed, impassable limit were dominant; just a couple decades later though, the Italian side adopted the same idea of the frontier. Despite the gradual opening of the border in the following decades, Piccinato's plan for Gorizia denied the existence of a modern city on the other side of the border. Once again, local needs regarding urban development were subordinated to the perception of the border as a delineator of separation.

Characteristics of the border changed as it gradually became passable and finally disappeared in the process of the enlargement of the EU and the Schengen area; these were reflected within the urban structure as an inner periphery, formed right in the heart of the conurbation extending over the two European countries. Through the absurdity of the traditional center-margins and interior-exterior dichotomies, the impact of the characteristics of the frontier on urban structures is revealed, that is, the frontier zone as the zone of separation and stoppage at a certain point in time, but the zone of linking and of passage in another era. Yet the new border influenced not only built structures, although without a doubt it created spaces that claimed certain new ways of living and functioning. It created a frontier zone, a space of open perception; for a time, it was a space of separation between two nations, cultures, histories and identities and at another time a space that unites diversity in a unique transnational conurbation.

REFERENCES

Angelillo, A. (ed.) 2004. *Gorizia-Nova Gorica. Progettare il confine*. Gorica: ACMA.

Angelillo, A., Angelillo, A., and Menato, C. (eds) 1994. *Città di confine, Conversazioni sul futuro di Gorizia e di Nova Gorica*. Portogruaro: Ediciclo Editore.

Anon. 1947. "O Gorici brez duše." *Soča: glasilo goriških, beneških in kanalskih Slovencev*, October 18, p. 3.

Bernik, S. 2004. *Slovenska arhitektura dvajsetega stoletja. Slovene Architecture of the Twentieth Century*. Ljubljana: Mestna galerija.

Bufon, M. 1995. *Prostor, meje, ljudje: razvoj prekomejnih odnosov, struktura obmejnega območja in vrednotenje dejavnosti na Goriškem*. Trst: Slovenski raziskovalni inštitut.

Cattunar, A. 2007. Gorizia, città di confine. La città 'immaginata' nei piani regolatori dal 1860 al 2008. *Studi di storia contemporanea*. Available at: https://www.academia.edu/7777071/Gorizia_citta_di_confine._La_citta_immaginata_nei_piani_regolatori_dal_1860_al_2008 (accessed July 2014).

Codellia, L. and Graziati, F. 2000. "La Nizza austriaca, la città redenta e la ricostruzione." In *Il Novecento a Gorizia. Ricerca di una identità. Urbanistica e architettura*. Venice: Marsilio, pp. 13–41.

Di Battista, A. 2011. "Nova odkritja pri raziskovanju Ravnikarjeve Nove Gorice in Trenzevega Velenja." *Zbornik za umetnostno zgodovino*, 47, pp. 322–40.

Fabiani, M. 1949a. "Castello e piazza della Vittoria." *Il Lunedì*, 18 April, p. 2.

Fabiani, M. 1949b. "Il nuovo orientamento di Gorizia. Il nostro castello." *Il Lunedì*, 11 April, p. 2.

Fabiani, M. 1949c. "Urbanistica Goriziana. Il rinnovamento della città." *Il Lunedi*, 28 March, p. 2.

Gabrič, A. 2005. "Osvoboditev in vzpostavitev nove oblasti." In Borak, N., Čepič, Z., Fischer, J. et al. 2005. *Slovenska novejša zgodovina. Od programa Zedinjena Slovenija do mednarodnega priznanja Republike Slovenije: 1848–1992*. Ljubljana: Mladinska knjiga; Inštitut za novejšo zgodovino, pp. 831–6.

"Gorica." In Javornik, M., ed. 1989. *Enciklopedija Slovenije, 3*. Ljubljana: Mladinska knjiga. pp. 279–82.

Ivanšek, F. (ed.) 1995. *Hommage à Edvard Ravnikar: 1907–1993*. Ljubljana: Self-published, France Ivanšek and Marta Ivanšek.

Jurca, N. 1990. "Travma meje. Obet Velike Gorice." *Urbani izziv*, 14, pp. 56–8.

Jurca, N. 1997. "Un piano per le due città. Pianificazione coordinata del sistema viario di Nova Gorica e di Gorizia." *Nuova Iniziativa Isontina*, 16, pp. 30–35.

Kopač, V. 1995. "Edo Ravnikar, risar in grafik v vojni in obnovi." In Ivanšek, F. (ed.) 1995. *Hommage à Edvard Ravnikar: 1907–1993*. Ljubljana: Self-published, France Ivanšek and Marta Ivanšek, pp. 212–15.

Koselj, N. 1995. *Arhitektura 60-ih let v Sloveniji: kontinuiteta ideje*. Ljubljana: Arhitektov bilten (special issue).

Krečič, P., Mušič, M., and Zupan, G. 1995. *Edvard Ravnikar: arhitekt, urbanist, oblikovalec teoretik, univerzitetni učitelj, publicist. Umetnostnozgodovinski oris*. Ljubljana: Arhitekturni muzej.

Malnič, A. 1998. "Topografija spomina na novo mejo." *Acta Histriae*, 6, pp. 331–46.

Marin, A. 2007. *Gorizia: piani e progetti per una città di confine.* Udine: Casamassima.

Morris, R.C. 2000. "The Counts of Gorizia Brought to Life." *The New York Times*, June 24. Available at: http://www.nytimes.com/2000/06/24/style/24iht-rod.2.t.html (accessed 2 August 2013).

Mušič, M. and Ivanšek, F. (eds) 1948. *Zbornik oddelka za arhitekturo na Univerzi v Ljubljani 1946–47.* Ljubljana: Državna založba Slovenije.

Nanut, C. 1972. "Ob 25-letnici najmlajšega mesta ob naši zahodni meji." *Srečanja. Revija za družbena vprašanja*, 35–36, pp. 7–14.

Nova Gorica. In Voglar, D. (ed.) 1994. *Enciklopedija Slovenije*, 8. Ljubljana: Mladinska knjiga, pp. 12–16.

Plahuta, S. 1979. "Pričetki gradnje Nove Gorice." *Jadranski koledar*, pp. 148–56.

Pozzetto, M. 2004. "Il laboratorio ACMA." In Angelillo, A. (ed.) 2004. *Gorizia–Nova Gorica. Progettare il confine.* Gorica: ACMA.

Progettare il confine per un territorio senza barriere: Convegno di studio e mostra dei progetti proposti perl'area di Gorizia e Nova Gorica al concosro internazionale di architettura EUROPAN 2. Načrtovanje meje za prostor brez barier: študijski posvet in razstave projektov za območje Gorica in Nova Gorica, ki so sodelovali na mednarodnem natečaju EUROPAN 2. Gorica and Nova Gorica, 1991.

Progetto di riconciliazione tra abitanti di Gorizia e di Nova Gorica, Gruppo di lavoro: urbanistica. Gorica and Nova Gorica, 1998.

Ravnikar, E. 1984. "Nova Gorica po 35 letih." *Arhitektov bilten*, 68–69, pp. 43–6.

Širok, K. 2010. "Spomin in pozaba na obmejnem območju: predstave o Goriški preteklosti." *Acta Histriae*, 1–2, pp. 337–58.

Spagna, E. 1973. "Sviluppo urbanistico a Gorizia ed esperienze di pianificazione." *Iniziativa isontina*, 2, pp. 15–24.

Spagna, E. 2003. "Gorizia Nova Gorica e Šempeter Vrtojba. Le nuove frontiere della pianificazione." *Rassegna tecnica del Friuli Venezia Giulia*, 54, pp. 10–16.

Šušmelj, J. 1997. "Odpiranje meje. Sodelovanje med Novo Gorico in Gorico." In Zoltan, J. (ed.) 1997. *Nova Gorica–Gorica: izzivi in možnosti sobivanja.* Nova Gorica: Mestna občina, pp. 9–28.

Torkar, V. 1990. "Dve Gorici—eno mesto. Ena Gorica—dve mesti." *Urbani izziv*, 14, pp. 51–2.

Torkar, V. 1990. "Racionalizacija urbane strukture Nove Gorice z arhitekturo mesta." *Arhitektov bilten*, 97–98, pp. 3–24.

Torkar, V. 1997. "Urbanizem in politika—primer Nova Gorica." In Valentinčič, S. (ed.) 1997. *Zbornik Primorske-50 let.* Koper: Primorske novice, pp. 15–18.

Torkar, V. 1999. "Quando un Confine genera una Città." *Architetti regione*, 27, pp. 10–11.

Troha, N. 2005. "Urejanje mejnih vprašanj." In Borak, N., Čepič, Z., Fischer, J. et al. 2005. *Slovenska novejša zgodovina. Od programa Zedinjena Slovenija do mednarodnega priznanja Republike Slovenije: 1848–1992.* Ljubljana: Mladinska knjiga; Inštitut za novejšo zgodovino, pp. 908–15.

Urbanistica e edilizia. In Bovio, C. (ed.) 1986. *Enciclopedia legale di Selezione. Tutta la legge dalla A alla Z.* Milano: Selezione R.D., pp. 967–8.

Vodopivec, A. and Žnidaršič, R. (eds) 2010. *Edvard Ravnikar: Architect and Teacher.* Wien/New York: Springer.

Vrišer, I. 1959. "Nastanek in razvoj Nove Gorice." *Geografski vestnik*, 31, pp. 45–78.

Vuga, T. 1990. "Združitev obeh Goric kot posledica urbanega razvoja." *Urbani izziv*, 14, pp. 59–62.

Zanini, P. 1997. *Significati del confine. I limiti naturali, storici, mentali*. Milano: Bruno Mandadori.

Žnidaršič, R. 2004. "Metoda projektiranja arhitekta Edvarda Ravnikarja." *Arhitektov bilten*, 165–166, pp. 8–33.

Zoltan, J. (ed.) 1997. *Nova Gorica–Gorica: izzivi in možnosti sobivanja*. Nova Gorica: Mestna občina.

10

New Urban Frontiers: Periurbanization and (Re)territorialization in Southeast Asia

Michael Leaf

URBANISM AND PERIURBANISM IN SOUTHEAST ASIA

Despite the tremendously increased speed and scale of global urbanization since the end of the Second World War, the world is still in the midst of its urban transition. Those labeled as the developed countries of the rich world have essentially completed their shifts from rural to urban, with 75 to 80 percent or more of their citizens now living in urban areas. Yet much of the developing world is projected to continue large-scale urban growth for the next few decades, so much so as to more than double both the number of urban dwellers and the total amount of urbanized territory on the planet between now and the middle of this century (Angel, 2012). For the developing, urbanizing countries of Southeast Asia much of this shift is occurring across wide swaths of territory on the peripheries of major metropolitan regions. One anticipated outcome of these processes of periurbanization is the formation of extensive metropolitan regions with multiple millions of residents. Because of their scale and speed, such changes are without precedent, thus leaving us with little basis for envisioning the nature of urban life in these places in the future. Such territories are clearly transitional, but without historical precedent for anticipating this sort of urban future, it is difficult to say what this might be a transition to.

My concern here is for the teleology of theory, as one may discern a basic split in the academic literature between one view that emphasizes the convergence of global urbanization patterns over time and another argument that emphasizes the special circumstances, if not uniqueness, of processes and patterns of urbanization in this region. Are Southeast Asian cities becoming more like those of the developed world, or are locally specific characteristics such as the formation and persistence of "desakota" zones[1] leading to new forms of urbanism? And if it is the latter, what are the characteristics that define this new form of urbanism?

In his seminal contribution to this debate, McGee (1991) emphasized such factors as high rural population densities in wet rice agricultural regions surrounding large cities, the shedding of labor from agricultural production in the wake of green revolution technological and institutional changes, and the rapid expansion of communications and transportation technologies across expanses of territory as underpinning the sort of transactional changes supportive of rapid livelihood diversification. In short, he identified a range of intersecting demographic, technological, economic and cultural shifts that fundamentally challenge the facile differentiation between urban and rural and lay the groundwork for rapid, regionally-based forms of urban transformation. In contrast to the expectations of convergence thinking, the new forms of urbanism that result from these processes may be as much about the persistence of rural characteristics and relationships as they are about societies becoming urban.

One important attribute of periurban change that is commonly emphasized is that of increasing fragmentation (Browder et al., 1995). Most obviously, this is seen in the outcomes of *ad hoc* and uneven land use conversion at multiple levels of scale, from clearly bounded (and gated) residential compounds, large and small, and clearly bounded (and gated) industrial development zones, to smaller scale residential and industrial sites developed through much more atomized or individualized plot-by-plot change. Using the label "fragmented" to describe the multiply-leapfrogged and multiply-scaled landscapes which result exposes a normative concern, with the implication that better "planning" (that is, the analysis of implications of change and the provision of appropriate infrastructure and institutions in advance of development, typically under state guidance or control) could serve to counteract the deleterious environmental and social effects of fragmented periurban spaces. In practice, however, even those components of Southeast Asia's periurban regions that are ostensibly "planned," meaning designed by or for the private development industry, can simultaneously be "unplanned," if they nonetheless contradict government land use plans.[2]

One tendency is to interpret this in terms of a weak state framework, with the need for institutional strengthening or "capacity building" in order to "formalize the informal," or otherwise to get a handle on the fragmented socio-spatial changes that seem to be so out of control. Fundamentally, then, this is about planning, or how the underpinning institutional relations (both ostensibly formal and informal) interact in shaping (peri)urban outcomes. In order to explore this further, I turn to the notion of the frontier, specifically to the idea that in Southeast Asian

1 This term was coined by McGee (1991; 1995) from Indonesian words for village (*desa*) and city (*kota*), and is used in reference to periurban territories that are seen to contain a mix of rural and urban functions.

2 Perhaps the single largest example of this in the region is the 6,000 ha "new town," Bumi Serpong Damai, developed since the late 1980s on the western edge of the Jakarta Metropolitan Region (Douglass, 1991).

periurban regions we are seeing the emergence of an urban frontier, or in actuality, multiple frontiers, when one considers the range of interactions shaping periurban outcomes.

THE MULTIPLE FRONTIERS OF PERIURBAN SOUTHEAST ASIA

Conventionally, a frontier is seen as a type of interface, and perhaps an "empty space" to be claimed and filled. It is also a place of encounter, of interaction and contestation between disparate groups, with the potential for new forms of social mixing, a place of promiscuity (Harms, 2006). But the frontier is also a discourse, implying newness and change. In this sense it is a place of hope, perhaps inevitability, and a source of worry and uncertainty; in many ways it is a metaphor for development. In much of the writing on frontiers in various contexts one also encounters the idea of lawlessness, with recourse to brute force as a principal means for the expression of power; from this we may understand institutional and regulatory weakness to be a fundamental characteristic of the frontier. Gaps open up, with ambiguity as to how they are to be filled. In the frontier's lawlessness, we may also see indications of its position as a geopolitical strategy, expressive of state interests and perhaps conditioned as well by market relationships. The processes of frontier formation are thus forms of territorialization, that is, the territorial expression of state intentionality (Vandergeest and Peluso, 1995). In the case of the rural/urban interface, one could argue that this is a reterritorialization, in reference to the reconfiguration of state-society relations as erstwhile rural zones become reclassified as urban.

Southeast Asian periurban zones may be thought of as frontiers in at least three senses. Most obviously, they are frontiers of urbanization in the broadest meaning of the word, resulting from the outward expansion of what are conventionally labeled as urban functions across ever-increasing territories. One would need to deconstruct the urban into its component parts in order to fully explore the implications of the periurban as the urban frontier. One finds, for example, the rapid growth of populations in such places (the demographic understanding of the urban) with increasing propinquity between suburbanizing urbanites, local former villagers and post-peasant migrants from perhaps much further away (Leaf, 2002). One also sees the rapid proliferation of non-agricultural activities and their functional interrelations, core components of what it means to be urban in economic terms. And one presumes periurban territories are also undergoing socio-cultural transformations,[3] though this may be less obvious to the outside observer.

The periurban is also a frontier of globalization, a point reiterated in the analytical literature that emphasizes the effects of global flows into these territories, specifically the flows of foreign direct investment (Webster, 2002). Under state policy regimes

---

[3] The demographic, economic and socio-cultural dimensions of urban change are what Friedmann (2002) refers to as the "three meanings" of urbanization.

heavily geared toward national development through export-oriented production, what were once the agrarian hinterlands of large cities have now been transformed into their industrial hinterlands as well. And to the degree to which these cities are also nodes of global connectivity for export flows, these should also be seen as real or aspiring industrial hinterlands of the global economy. Yet globalization is not only manifested through the material flows of capital and goods. It also permeates these territories through symbolic and ideational flows, from the media-conveyed imagery of newness and modernity that inspires the desires of urban-turning migrants from ever further afield, to the stylistic cues and accompanying marketing strategies of city-based real estate development companies working in these post-rural territories. In a sense, the rapid outward expansion of the urban spatial economy, commonly denounced as land speculation, is arguably also an aspect of the periurban as a frontier of globalization, as here one sees the ascendancy of market logic as the principle mechanism of allocation—the neoliberalization, as it were, of emergent metropolitan regions. To the extent that such territories are becoming directly linked into globally-integrated property markets, for residential real estate or for industrial locations or even for new recreational landscapes, these are new territories of global connectivity.

There is a third and perhaps more critical aspect of the periurban as a frontier, which is what I would refer to as a governance frontier. This is the result of ongoing reterritorialization by the state, particularly, though not exclusively, the local state. At the level of formal institutions this tendency has been apparent for some time in the establishment and attempted strengthening of cooperative regional planning frameworks, such as Jabotabek (now Jabodetabek) in Jakarta, or Calabarzon for Manila, or the Bangkok Metropolitan Region. In more explicit administrative terms, this is the rationale underpinning urban boundary redefinition, whether carried out incrementally, as with the various new urban districts carved out of Ho Chi Minh City's surrounding rural territories in recent years, or through the sort of wholesale annexation of nearby provinces as carried out by the Hanoi municipal government in August 2008.[4] The idea of the periurban as a frontier of governance driven by processes of reterritorialization, however, is more complex than what is implied by the ostensible redrawing of boundary lines. Embedded in these trends are also the intrinsic expressions of regulatory power, possibly indicating contestation at levels of territorial elites (Malesky, 2006), as well as configuring the decision-making underpinnings of "everyday urbanization" (Kelly, 1999). The reterritorialization referred to here, then, is not purely an administrative matter, as it bleeds over into questions of property rights and even more fundamentally into whether the state is able to draw upon sufficient regulatory power to overcome alternative pressures

4 "Greater Hanoi" was designated as of August 1 to include the capital city and what was formerly 219,000 ha of Ha Tay province, Vinh Phuc province's Me Linh district and four communes in Hoa Binh province's Luong Son district. The expansion increased the total territory of the municipality/province of Hanoi from 92,000 ha to nearly 334,500 ha, with a consequent increase in official population figures from 3.5 million to over 6.2 million.

from either embedded local land disposition practices on the one hand or from the urban-based development industry on the other. How such questions play out in any context will be determined by the particularities of local political economies of urbanization; in the following section, I turn to the context of Ho Chi Minh City by way of example.

CONTEXTUALIZING HO CHI MINH CITY'S (PERI)URBAN FRONTIER(S)

Ho Chi Minh City (HCMC), Vietnam's largest urbanized area and principal commercial centre, has undergone tremendous changes in recent decades. Looking at what is perhaps the single most basic demographic indicator of urban change, we see that the city's overall population has grown rapidly since the early days of the *doi moi* ("renovation") reform period (as discussed below). In official figures, the total population (nominally urban and rural districts combined) has gone from just over 4.1 million in 1990 to nearly 6.7 million in 2007, or an increase of over 61 percent in just 17 years (HCMC Statistical Office 2008, p. 24). Yet official figures are notoriously inaccurate. One estimate, for example, placed the total population at 7.5 million in 1998, at a time when the official figure was closer to 5.1 million (Gainsborough, 2003, p. 3, fn 18), showing a variance to official numbers of nearly 50 percent. Undoubtedly, a large part of this discrepancy derives from the institution of Vietnam's household registration (*ho khau*) system with the major portion of the extra couple of million inhabitants in the city being "people out of place," rural migrants from outside the city and hence off the official record books, at least to the extent that these numbers are aggregated to the municipal level.

This issue of aggregation of statistical information is more consequential than most casual observers would think, as it gets to the heart of the question of who maintains control, or at least control of information, and at which level of the governmental hierarchy. Clearly the extra 2.4 million inhabitants of the city (as of 1998, undoubtedly higher today) are neither missing nor hiding. A visit to any ward office shows that local officials know where these people are and, indeed, a nationally implemented accounting system has been created to organize and presumably keep track of population movements. Under this scheme, everyone is grouped into one of four categories depending on the correspondence (or lack thereof) between the address on a person's *ho khau* and where they actually live. Thus everyone is labeled as either KT1[5] (those whose addresses match), KT2 (those whose addresses don't match but are both within the same province), KT3 (those from outside the province but are considered to be settled in their new locales), or KT4 (out-of-province migrants whose status is considered temporary). These categories are used in the maintenance and continual upgrading of records at ward offices. Undoubtedly there are some who fall through the cracks in the KT system, yet we can be assured that officials on the ground certainly know what is going on within their administrative territories.

---

5 An abbreviation of *kiem tra*, meaning "control" or "inspect."

A further contextualizing factor shaping HCMC's periurbanization is the *doi moi* policy regime, ostensibly put in place in response to the economic difficulties under the planned economy of a unified Vietnam in the 1980s, and characterized by an open door approach and a move toward establishing market principles as the basic allocative mechanism for societal goods. Yet there is the concern here as to whether the "reform" of the *doi moi* period should be seen to refer to the formal policy regime *per se*, or more broadly to the shifts in mores and practices across society. Is *doi moi* best viewed as a state initiative or as a state response to a societal phenomenon? (Fforde and de Vylder, 1996). Critical in either respect have been the effects of decentralization, as the Vietnamese party-state has sought to establish a balance between the devolved autonomy seen as necessary for underpinning the move away from "democratic centralism" and the practices of a command economy, and the need for maintaining legitimacy for central leadership. In his analysis of the political economy of HCMC over the course of the 1990s, Gainsborough (2003) emphasizes the challenges of attaining this balance, and stresses in particular how the opening up and devolution of authority in the reform period have fostered parallel and highly consequential informal patterns of collaboration and decision-making linking party, state and private interests at local levels.

One is hesitant, though, to see this in dichotomous terms as a formal/informal difference, as in practice such distinctions are not clearly demarcated and the actual functioning of formal structures may in many ways determine the unfolding of ostensibly informal strategies. This point—and in particular how interlocking authority structures, from the central government, to the municipality, to local ward offices, shape the development of HCMC's periurban areas—will be illustrated in the following sections in reference to specific elements of HCMC's periurbanization, residential development and industrial development. Aspects of this correspond roughly to the first two "frontiers" identified above. Following this, I end with a discussion of the implications for periurban administrative change, or the question of the evolving governance frontier.

RESIDENTIAL DEVELOPMENT IN PERIURBAN HCMC

In contrast to the current emphasis in western planning practice on "compact cities," HCMC has seen rapid deconcentration of older inner city districts and expansion in surrounding areas. As would be expected in such a context of growing socio-spatial differentiation, newly constructed housing takes a wide range of forms. The residential development project that has garnered the most attention is Saigon South (Nam Sai Gon) and within it the development of Phu My Hung.[6] This project, or interlinked set of projects, has become emblematic of

6 Saigon South covers a territory of 3,300ha and is planned to house a population of 500,000, with an additional daytime population of another 500,000, when it is completed in 2020. Phu My Hung Corporation, a land development subsidiary of CT&D, was responsible for arranging the masterplanning of the entire Saigon South

10.1 The Nguyen Van Linh Expressway and the entrance to Phu My Hung (Saigon South).

HCMC's efforts to globalize, to recast itself as a player on the world stage. Saigon South is portrayed in academic literature as an important part of the city's self-conscious attempt at "world city formation," a privatized urban "utopia" created through an internationally-organized mega-project (Waibel, 2004, 2006; Douglass and Huang, 2007). The project itself covers a wide band of territory along the city's southern border, following an east-west corridor created by the construction of the Nguyen Van Linh Expressway (Figure 10.1), which illustrates the importance of infrastructural development for opening new zones of periurban expansion. The Phu My Hung development offers a range of building types, from a highly protected internal core enclave of extremely high-end "villas" to outer blocks of more modest high-rise apartments with a built fabric of mid-rise urban apartment blocks (Figure 10.2) in between.

For our purposes, however, we should understand the story of Saigon South as pre-dating the project itself, rooted in geopolitics, and shaped by efforts at central state developmentalism in the early reform period. The deep roots of Saigon South and Phu My Hung may be traced from the early 1990s, through an alliance of interests between two governing parties, the Communist Party of Vietnam and the Nationalist Party (KMT) of the Republic of China (Taiwan). Though seemingly strange bedfellows, there were complementary interests at play: the Taiwanese wanted to diversify their economy through a strategy of internationalization, while

project area, though it is the developer of only the easternmost 20 percent of the site, administratively consisting of two wards of District 7.

10.2 An apartment block in Phu My Hung.

the Vietnamese were looking to open up their economy in a manner that was ideologically acceptable in the early days of *doimoi*, with Taiwan seen as neither a western nor a communist partner.

The Taiwanese corporate partner, Central Trading and Development (CT&D), was originally an investment vehicle of the KMT, although the party had divested itself of it by 1993. The company's first Vietnamese project was in forestry in Kien Giang province, before it was invited to develop the Tan Thuan EPZ, Vietnam's first industrial park and export processing zone, located in District 7, downriver from the old core of HCMC. As a step toward securing Tan Thuan as an internationally competitive industrial zone, CT&D also undertook the development of the city's first large-scale power plant, Hiep Phuoc, so as to have a continuous source of power to the EPZ. Considering this "prehistory" of Saigon South, one could argue that this periurban megaproject is only ambiguously a municipal strategy toward globalization; more broadly it is consonant with central government efforts to develop the nation through international capital and technology transfers. Though often presented as an exemplar of housing development in HCMC, Saigon South should instead be seen as a special case, growing out of past "pioneering" projects that established CT&D as a committed long-term partner of the Vietnamese government in their modernization strategy.

Yet there is no single model of residential land development on the edge of HCMC. Even within the luxury end of the market, there are critical distinctions between the gated compounds one finds within projects such as Phu My Hung, and the plot-by-plot development of single houses in desirable areas such as Thao Dien and An Phu to the northeast of the city. And it is striking how the business of real estate has developed so quickly over the course of the *doi moi* period (Kim 2008), fostered as it has been by the HCMC municipal government as an income-generating activity for the city. Emblematic of this aspect of the industry is the development firm Intresco (Investment and Trading of Real Estate Joint Stock Company). This company, now in its third decade, was set up as a state-run company and equitized (opened to non-state capital) in 2001, although 20 percent is still held by the government. Originally established as a vehicle for the city government to engage in real estate development of various sorts, its portfolio was fairly small until it was equitized, at which point it was able to expand quickly. It is now one of the top 500 companies in Vietnam, drawing capitalization from more than 600 stockholders, including the major foreign investment funds that operate in Vietnam (Mekong Capital, Vina Capital, and others). The company is thus able to pursue a range of both commercial and residential development projects including, among others, sites within Saigon South (Figure 10.3).

Residential real estate is big business; it is not just a process to create housing, but a means for both investors and state actors to make money. For the developers and buyers in the formal real estate business, this has been an extraordinarily lucrative undertaking. This is perhaps an obvious statement for the burgeoning though highly competitive luxury and upper-middle class markets in which Intresco and Phu My Hung operate. It may be equally true, however, for those who provide housing for the myriad rural-urban migrants on the edge of the city, at the bottom end of an enormous and highly segmented residential market.

Foreigners often remark how striking it is that HCMC has few squatter settlements in comparison to other major cities in Southeast Asia. The reason for this may best be attributed to the particular structure of implicit property rights held over from before the reform period, when villagers at the edge of the city were given latitude to determine the uses of their residential lands. Over the years, as the market economy has taken hold, the edge of the city has industrialized, and migrants have shown up in increasing quantities, erstwhile villagers have shifted out of agriculture to become small-scale landlords by building "lodging houses" (*nha tro*) on their residential lands (Figure 10.4). Typically, these consist of a series of basic cinder-block rooms of perhaps 10 or 12 square meters each, rented out to migrants on a monthly basis (Phan 2007). From figures obtained in one recently urbanized village on the western edge of HCMC, we find that 553 village households (meaning KT1), or approximately 20 percent of the total, had registered as owners of *nha tro*. Within this village, the number of migrants (KT4) who are housed in this way (12,500) is nearly equal to the total number of KT1 villagers (13,000) with, on average, each landlord household providing accommodations for approximately 22 migrants.[7]

[7] This is in addition to the 4,500 KT3 migrants who had been able to buy their own houses or houseplots, and 9,900 KT2 urbanites who had moved in from elsewhere in the city, thus comprising a total registered population for the former village (and new ward) of about 40,000.

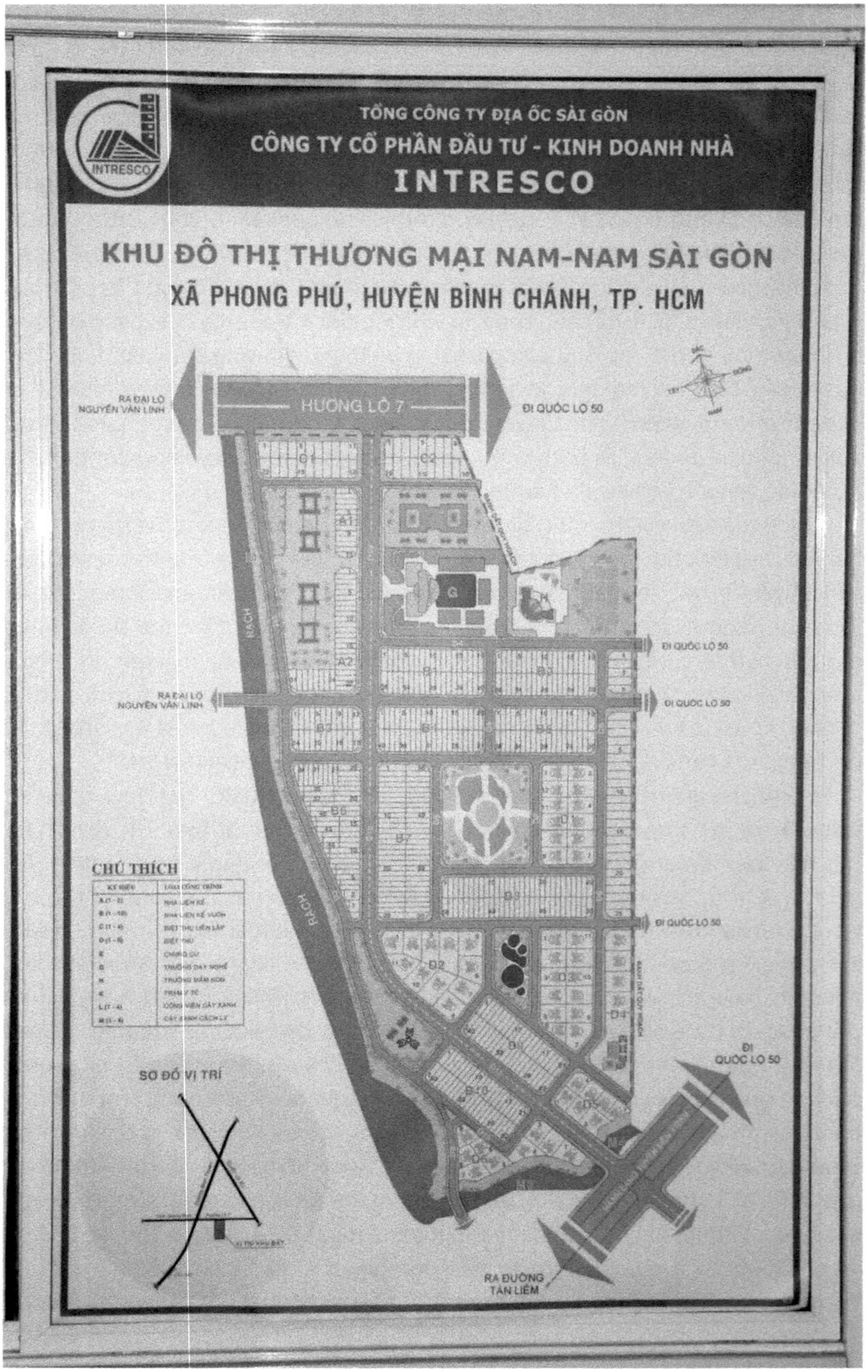

10.3 Site plan for a project undertaken by Intresco in Saigon South.

10.4 Shared entranceway for a series of *nha tro* (lodging house) rental units.

This one example among many on the urbanizing edge of the city raises the intriguing question of why a form of housing that is nominally illegal is not only tolerated by local officials (the *nha tro* and their KT4 residents are registered at the ward office, after all) but seemingly even encouraged. To understand this, one can turn to an argument regarding local state legitimacy, particularly in the context of devolution of powers and responsibilities to lower levels in the state hierarchy. In the current policy context of decentralization, there are increasing fiscal pressures put on district-level (and by extension, ward-level) offices, with an almost competitive spirit among local officials to generate as much revenue for municipal coffers as possible in the interest of maximizing redistributions back to the locality. In light of an overarching economic development strategy that stresses rapid industrialization and that utilizes the low-cost labor of rural-urban migrants as a means to remain competitive, the informal solution of allowing the proliferation of *nhatro* housing makes a good deal of sense. By providing "low-cost" (though substandard) housing, the reproduction costs of labor are kept low, thus increasing the competitiveness of localities in attracting private industrial investment and in turn strengthening local state revenues through the collection of taxes and fees. And allowing formerly agricultural villagers to do this extends the benefits to local constituents, specifically the KT1 residents for whom local officials are primarily responsible. Thus a coalition of interests is formed, with local residents benefiting through the rents they receive, migrants benefiting from the low-cost housing they are able to access, factory owners benefiting from the low-cost labor they obtain, local officials benefiting from the revenues they are able to generate locally, and higher level officials benefiting from generated wealth for the city overall.

Through this strategy of regulatory flexibility, the urban fortunes to be derived from development of the city's edge are thus utilized to address a diversity of interests. And when one considers the seemingly chaotic overall picture of periurban residential development, this appears as one component of an elaborated ecology of interests, involving state and non-state actors both domestic and foreign, and at a range of scales from the centre down to the ward and even to the individual households within the ward. Rather than seeing HCMC's periurban landscape as highly fragmented, with housing developments categorized as either formal or informal, one may instead start to understand this as an integrated, though highly segmented, system arising from a coalescence of state and non-state interests and with the involvement of foreign capital at various levels of scale, from the central state-supported cooperation underpinning Saigon South, to the equitization of firms such as Intresco, to the interests of off-book foreign buyers and renters of housing in places such as An Phu, to the smaller-scale industrial investors whose factories dot the periurban landscape and underpin the viability of *nha tro* housing for the legions of new migrants to the city.

INDUSTRIAL DEVELOPMENT IN PERIURBAN HCMC

A look at the seemingly disparate industrial landscape that has developed in periurban HCMC reveals a scene similar to that of housing. And for that matter, one could as well begin the story with Taiwan's CT&D and their close working relationship with Vietnam's central government. On the ostensibly formal side of things, HCMC's industrialization has been propelled through the establishment of what are currently 12 industrial parks (*khu cong nghiep*) and three export processing zones (*khu che xuat*), which provide a major vehicle for foreign investment and can be seen as a significant platform for Vietnam's labor transition from agriculture to higher value-added forms of industrial production. These zones are under the administration of a state agency known as HEPZA (HCMC Export Processing and Industrial Zones Authority). Overall they have been highly successful, providing sites for foreign firms to establish their factories and offices (Figure 10.5) and thereby attracting more than USD 4 billion in investments by 2008 and creating more than a quarter of a million new jobs in the city.

HEPZA itself was established in 1992, following the opening of Vietnam's first EPZ, CT&D's jointly-developed Tan Thuan EPZ, in 1991. Although the "H" in HEPZA stands for Ho Chi Minh City, the agency is not a component of the municipal government but instead reports directly to the Prime Minister's Office; as a "special model" it is also independent of the various line ministries of the central government that might otherwise be assumed to have an interest in the development of the city. Like Saigon South in the area of residential development, the Tan Thuan EPZ is Vietnam's exemplar for industrialization, the model to be emulated in the development of HEPZA's subsequent suite of projects. And although seemingly an

10.5 Office of a Taiwanese-invested manufacturing business within an industrial park.

important part of HCMC's economic development strategy, HEPZA provides both the mechanism for central government oversight as well as a means for maintaining some autonomy from both local and higher-level ministerial structures.

Even a brief tour through HCMC's periurban territories exposes the extent to which HEPZA's 15 industrial and export processing zones are by no means the only game in town when it comes to the city's ongoing industrialization. To take the example of the recently-established ward referred to above, this territory with a total population of around 40,000 contains within it more than 200 locally-registered factories, in addition to more than 1,500 household-based small businesses. On a journey down the ward's main roads one sees steel-rolling mills, plastics factories, and all manner of production facilities interspersed with houses large and small, motorcycle repair shops and a range of roadside businesses (Figure 10.6). One also finds a noticeable presence of foreign investment, as typically the factories are joint ventures with partners from Taiwan, Singapore, Malaysia and elsewhere, a form of investment referred to in a Chinese context by one observer as "guerrilla" investment (Hsing, 1998), noteworthy for being small-scale and dispersed although significant in aggregate. Even if such factories fall into a regulatory grey zone, however, they may nonetheless be built to a standard of quality not too different from those in the formal industrial zones (Figure 10.7).

10.6 A Taiwanese-invested manufacturing business located outside an industrial park.

This mixed, industrial landscape can be understood as a correlate of sorts to the widely diverse production of housing described above, with the establishment of periurban factories negotiated and registered through local ward offices. And as with the seemingly "informal" *nha tro* housing, which is also registered at local ward offices, it is nearly impossible to form an aggregate picture of this situation for the city overall. Like the figures for the city's population and housing stock, HCMC's official statistics for such things as numbers of factories, industrial output, foreign industrial investment and industrial labor force are undoubtedly highly inaccurate, with local planners and researchers estimating (or in all probability only guessing) that there may be as much or more industrial production outside the established industrial estates as within them.

In this paper so far I have divided questions of periurban industrial development from questions of residential expansion. Clearly, though, the two are intimately interlinked, with industrialization and housing production at the bottom end of the scale perhaps best understood as an integrated system, as guerrilla factories provide employment and *nha tro* landlords provide the housing for the burgeoning numbers of newly (peri)urbanizing rural migrants. Considering the tremendous scale and economic significance of these ongoing changes, it is curious that so much of this appears to be "off the books" or seemingly unregulated by higher levels of authority. For this, it is useful to return to the theme of periurbanization as a governance frontier in the concluding section.

10.7 Garment manufacturing in an independent factory (outside of an industrial park) with Hong Kong and Canadian investment.

PERIURBAN HCMC AS A GOVERNANCE FRONTIER

In nominal terms, the reterritorialization of periurban HCMC may be taken to refer to the redesignation of officially rural districts as urban districts, or in other words, the exercise of redrawing the city's urban boundaries. In the Vietnamese administrative structure, sub-provincial districts are of two types, referred to as *huyen* for what are designated as rural areas, and *quan* for their urban counterparts. The provincial-level municipality (*thanh pho*) of HCMC, for example, is currently comprised of 19 *quan* and 5 *huyen*. Redistricting and redesignation have occurred on two occasions in recent years. First, in 1997, one *huyen* across the Saigon River to the east of the city core (Huyen Thu Duc) was broken into three new *quan* (Quan 2, Quan 9 and Quan Thu Duc), and two more *quan* were carved out of other *huyen* (Quan 7 from Huyen Nha Be and Quan 12 from Huyen Hoc Mon).[8] And again, in 2003, a new *quan* (Quan Binh Tan) was carved out of Huyen Binh Chanh, on the western edge of the city.

The designations of *huyen* and *quan*, though nominally referring to rural and urban districts, only roughly correspond to shifting patterns of rural and urban land use. The territory of what is now Saigon South covers parts of Huyen Binh Chanh,

---

[8] The curious mix of both numerical and historic place names for the various *quan* is owing to the history of what is now Ho Chi Minh City, as the municipality is comprised of two older cities, Sai Gon (whose *quan* were numbered) and Gia Dinh (whose *quan* were given historic place names).

a section of the recently designated Quan 7, and parts of the still nominally rural Huyen Nha Be. On the other hand, the recently established Quan 9 still contains within it large portions of agricultural land. From interviews with urban planning officials, it is apparent that there is no standardized procedure for guiding the redesignation of rural territories as urban, and that this administrative change may either precede the functional changes of urbanization across the rural landscape (as with Quan 9) or follow upon it (as with Quan Binh Tan). Considering the emphasis in HCMC's current master plan to build up new outlying urban sub-centres and to create *de novo* a satellite city in Huyen Cu Chi to the far northwest, one can expect that further administrative shifts will occur in coming years.

The interpretation of reterritorialization, however, is not limited to these official administrative changes, as it obliges us to consider changing authority structures, both formal and informal, as urban functions and the urban spatial economy intersect with rural practices and rural land uses. Here I would emphasize the frontier characteristic of lawlessness, or how the institutional unsettledness of these transitions opens up space for accommodation through regulatory flexibility. Such accommodation exists at the top end of the urban development pyramid with, for example, the establishment of a "one gate" agency such as that which exists to facilitate the involvement of investors in the city's industrial development zones. At the other end, accommodation is also reflective of local power relations as they influence the socio-economic conditions of urbanization, such as with the local registration and oversight of nominally illegal *nha tro* housing.

For zones in the midst of ongoing administrative shifts, decisions around such questions as land use conversion, the control of who may reside where, or what sorts of infrastructure and services are provided and by whom, may draw upon multiple and potentially conflicting authority structures. The fundamental question here is comparable to a long-standing area of inquiry in the literature on informal settlements: how is tenure secured in situations where urban construction and land occupation occur outside of officially prescribed regulatory frameworks? Briefly, the answer to this lies in the understanding that state authority, as locally constituted, may be insufficient relative to other sources of societal power, whether these derive from a local village council, a religious organization, a particularly well-connected land-owner, or perhaps even a local criminal gang (Leaf, 1994). In the case of the development of *nha tro* housing on the western edge of HCMC, the necessary accommodation for urbanization to occur happens through the negotiated, shared interests of local officials, village landlords and industrial investors.

I have loosely been using the contentious terms "formal" and "informal" in describing this situation, terms that are themselves theoretical referents to particular strands of literature. Despite tendencies among many writers to use these words dichotomously, here I would emphasize that the seemingly fragmented forms of urban change one finds on the edges of HCMC are best understood as an entire, integrated, though perhaps disarticulated system. The reterritorialization that one observes as the city expands outward responds to the interests of varying sets of actors in different sites and in various ways, from central party-state interests in conjunction with foreign developers and emergent middle-class property investors in a location such as Phu My Hung, to local villagers in cooperation with newly-designated ward officials for the development of *nha tro*. The idea that there is a sharp distinction between "formal" (as adhering to stipulated state norms and processes) and "informal" (as that which goes on in spite of the state) does not apply in this instance. The reterritorialization that occurs as HCMC extends outward is apparent in the negotiation of interests between a number of sets of actors, including the multiple layers of a decentralizing party-state system, multiple constituencies of social actors for whom the party-state must legitimize itself, and the growing importance of "market forces" throughout.

In thinking about market forces, the dichotomous distinction between "foreign" and "domestic" is also problematic in this context. As a frontier of globalization, transnational connectivities are interwoven throughout HCMC's periurban region, from the presence of major foreign players in the city's industrial zones and high-end property markets to the "guerrilla" industrial investors whose factories line the roads of the periurban. Thus the challenge of periurban governance—that is, articulating and balancing the potentially conflicting interests of state, society and capital—must necessarily engage questions of globalization and the attendant issues of local dependency on conditions beyond national borders. There is a danger here in that the positive benefits of the city's outward growth have for the most part occurred in a time of economic expansion, when the fortunes of urbanization could be shared widely among a diversity of actors, creating opportunities for urban elites and rural migrants alike. In the period of a potentially prolonged recession in the global economy, we should be concerned that the periurban frontier of Ho Chi Minh City may rapidly devolve into a landscape of anxiety and uncertainty. In this, we are reminded that unpredictability is apparently inherent to the notion of the frontier.

ACKNOWLEDGMENTS

Funding for carrying out the research on which this paper is based was provided by the Social Sciences and Humanities Research Council of Canada, for which I am grateful. I would also like to express my gratitude to Van Thi Ngoc Lan and Nguyen Quang Vinh of the Southern Institute for the Social Sciences for hosting my visit to Ho Chi Minh City and to Bui Duc Kinh for his untiring research assistance.

REFERENCES

Angel, S. 2012. *Planet of Cities*. Cambridge, MA: Lincoln Institute of Land Policy.

Browder, J.O., James R.B., and Joseph, L.S. 1995. "Patterns of development on the metropolitan fringe: Urban fringe expansion in Bangkok, Jakarta, and Santiago." *Journal of the American Planning Association*, 61(3), pp. 310–27.

Douglass, M. 1991. "Planning for environmental sustainability in the extended Jakarta Metropolitan Region." In N. Ginsburg, B. Koppel, and T.G. McGee (eds) 1991. *The Extended Metropolis: Settlement Transition in Asia*. Honolulu: University of Hawaii Press.

Douglass, M. and Huang, L., 2007. "Globalizing the city in Southeast Asia: Utopia on the edge—the case of Phu My Hung, Saigon." *International Journal of Asia-Pacific Studies*, 3(2), pp. 1–42.

Fforde, A. and de Vylder, S. 1996. *From Plan to Market: The Economic Transition in Vietnam*. Boulder, CO: Westview Press.

Friedmann, J. 2002. *The Prospect of Cities*. Minneapolis: University of Minnesota Press.

Gainsborough, M. 2003. *Changing Political Economy of Vietnam: The Case of Ho Chi Minh City*. London/New York: Routledge Curzon.

Harms, E.L. 2006. *Saigon's Edge: Transforming Time and Space on Ho Chi Minh City's Rural-Urban Margin*. Unpublished doctoral dissertation, Cornell University, Department of Anthropology.

HCMC Statistical Office. 2008. *Ho Chi Minh City Statistical Yearbook 2007*. Hanoi: Statistical Publishing House.

Hsing, Y. 1998. *Making Capitalism in China: The Taiwanese Connection*. New York: Oxford University Press.

Kelly, P.F. 1999. "Everyday urbanization: The social dynamics of development in Manila's extended metropolitan region." *International Journal of Urban and Regional Research*, 23(2) pp. 283–303.

Kim, A.M. 2008. *Learning to be Capitalists: Entrepreneurs in Vietnam's Transition Economy*. Oxford/New York: Oxford University Press.

Leaf, M. 1994. "Legal Authority in an Extralegal Setting: The Case of Land Rights in Jakarta, Indonesia." *Journal of Planning Education and Research*, Fall, pp. 12–18.

Leaf, M. 2002. "A Tale of Two Villages: Globalization and Peri-Urban Change in China and Vietnam." *Cities*, 19(1), pp. 23–31.

Malesky, E. 2006. "Gerrymandering—Vietnamese Style: The Political Motivations behind the Creation of New Provinces in Vietnam." Paper presented to the Annual Meeting of the American Political Science Association, Philadelphia, PA, September 3, 2006.

McGee, T.G. 1991. "The emergence of Desakota regions in Asia: Expanding a hypothesis." In N. Ginsburg, B. Koppel, and T.G. McGee (eds) 1991. *The Extended Metropolis: Settlement Transition in Asia*. Honolulu: University of Hawaii Press, pp. 3–26.

McGee, T.G. 1995. "Metrofitting the emerging mega-urban regions of ASEAN: An overview." In T.G. McGee and I.M. Robinson (eds) 1995. *The Mega-Urban Regions of Southeast Asia*. Vancouver: UBC Press, pp. 3–26.

Vandergeest, P. and Peluso N. 1995. "Territorialization and State Power in Thailand." *Theory and Society*, 24, pp. 385–426.

Waibel, M. 2004. "The development of Saigon South New Urban Area. A sign of an increasing internationalization and polarization in Vietnamese society." *Pacific News*, 22, pp. 10–13.

Waibel, M. 2006. "The Production of Urban Space in Vietnam's Metropolis in the course of Transition: Internationalization, Polarization and Newly Emerging Lifestyles in Vietnamese Society." *TRIALOG—A Journal for Planning and Building in the Third World*, 89(2), pp. 43–8.

Webster, D. 2002. "On the Edge: Shaping the Future of Peri-Urban Asia." Discussion paper, Asia/Pacific Research Center. Stanford: Stanford University.

11

Mediterranean Frontiers: Ontology of a Bounded Space in Crisis

Antonio Petrov

> *The spirit now wills his own will, and he who had been lost to the world now conquers the world.*[1]
>
> *Thus Spoke Zarathustra, Friedrich Nietzsche*

"What is the Mediterranean?" The *Oxford English Dictionary* responds to the question by defining "Mediterranean," from the Latin *mediterraneus*, meaning "in the middle of the land," as an adjective that characterizes an almost landlocked space. But it also suggests distinct characteristics and stereotypes that indicate the boundedness—characteristics and degree of boundaries seen from many different perspectives—of a territory, without actually determining how bound or unbound the territorial and cultural limits are. Cultural historian Iain Chambers defines this territory in-between Europe, Asia, and Africa as "Medi-terranean," arguing it is "an interleaved and multi-stratified constellation, a point of dispersion and dissemination, rather than a single, concentrated unity" (Chambers, 2013).

Needless to say, the Mediterranean is complicated and multifarious. Any attempt to capture the region seems implausible; no single definition or spatial determination can include its multivalent readings, cultivations, and mappings. It still is characterized by a tumultuous history and the ups and downs of civilizations that have flourished around the sea: Egyptians, Greeks, Romans, and Arabs, just to name a few, have defined the history of human civilization at this intersection of three continents. Whether as part of the cultural history of the world—with roots in classical civilizations of Greece, the Roman Empire, or Renaissance Italy; the religious history of the world—Judaism, Christianity, and Islam; or its economic history—with the Genoese, Venetians, or Catalans; the fundamental role of the Mediterranean in the development of global history cannot be denied (Abulafia, 2002).

---

1 Nietzsche, Friedrich Wilhelm, and R.J. Hollingdale. *Thus Spoke Zarathustra: A Book for Everyone and No One*. Penguin Classics. Harmondsworth, Middlesex, UK: Penguin Books, 1969.

11.1 Yona Friedman, "La Venise Monégaque". © 2014 Artists Rights Society (ARS), New York/ ADAGP, Paris

Mediterranean scholars—historians such as Fernand Braudel, Nicolas Purcell and Peregrine Horden, David Abulafia, or social scientists such as Henri Lefebvre and Michael Herzfeld whose sometimes overlapping and contradictory perspectives we continue to use—answer questions about the Mediterranean in their own way and employ different epistemic frameworks within which they believe the region unfolds (Petrov, 2013). The study of the Mediterranean is so rich that the methodological conjectures over investigations invite differences between disciplinary perspectives, as does the Mediterranean itself by emphasizing "the narcissism of small differences" keeping the omnipresence of distinctions within the territory in constant tension (Sarkis, Herzfeld, Ben-Yehoyada, 2013).

With characteristics so elusive, and a geographic coherence so ambiguous, this chapter examines how utopian projects like Herman Sörgel's *Atlantropa* and Rem Koolhaas' *Eneropa* do greater justice to a coherent perception of a Mediterranean territorial integrity by proposing new conceptions of the region's synthesis. I argue that by doing so, these projects not only serve as alternatives to scholarly readings of what the Mediterranean is, but also, one could say they do so by radically eliminating "the Mediterranean." Both schemes replace territorial structures that make the fabric of the Mediterranean and insert modern frameworks that overwrite existing borders, boundaries, and frontiers. Sörgel and Koolhaas promise new territories that not only are sites of new (modern) world orders, but also function as a pivot between geographical and utopian imaginations.

By examining history, geography, and identity as epistemic frameworks—all elements of transformation—I aim to recover how an imaginary of what the Mediterranean (potentially, or actually) is has imposed itself between reality and its symbolic image, upsetting the balance between the two. In a region in which an ambiguous geographic identity has been tenaciously preserved over millennia, in part through its architectural tropes and stereotypes, historical, geographic, and cultural definitions of a Mediterraneanity—causes and effects in relation to a larger Mediterranean identity—evoke imperceptible bonds beyond stereotypical manifestations that not only determine new regional formations, boundaries, and frontiers, but also distort any cultural *a priori*.

Are there ways to conceptualize the Mediterranean without "the Mediterranean?" And can the dichotomy between the consistent crisis of spatial and identity politics on one hand, and a Mediterraneanity driven by stereotypes on the other, produce new morphologies that consider complex geographic and cultural parameters of a present-day Mediterranean, if not the world?

HISTORY

> *To draw a boundary around anything is to define, analyze, and reconstruct it, in this case select, indeed adopt, a philosophy of history.*[2]
>
> *Fernand Braudel*

In *The Corrupting Sea*, Nicolas Purcell and Peregrine Horden take this fundamental but provocative question—"What is the Mediterranean?"—to new ends. The way they engage with the problem and how they answer it "is as radical as the method they use to arrive at it" (Lahoud, 2013). However, it is impossible to approach their text without looking at the canonical propositions that French historian and leader of the Annales School, Fernand Braudel, makes in *The Mediterranean and the Mediterranean World in the Age of Philip II*. Both books question the historic scale of the Mediterranean in new ways, but they depart radically from each other by taking fundamentally different methodological approaches. Rather than organizing history into separate layers as Braudel does, Horden and Purcell begin to define their problem by asking a question that suspended Braudel's initial proposition. Braudel stressed the analysis of structures and the endless stretching and layering of boundaries and common rhythms in the Mediterranean (Braudel, 1995). In contrast, Horden and Purcell highlight how "kaleidoscopic fragmentations," in which the tectonics of the region define "microregions," are based on topographical conditions that delineate how mutual characteristics are points of contention rather than common grounds (Horden and Purcell, 2000).

For Greek philosopher Plato, the Mediterranean is a set of common territories. In a famous passage of *Phaedo*, Plato envisions a Mediterranean world by describing

---

[2] Fernand Braudel, *The Mediterranean and the Mediterranean World in the Age of Philip II*, Sian Reynolds trans. (New York, 1972), 18.

a common territory in which Greeks gather around a pond like frogs. He describes the experience of a geographically captured space as part of a larger world with other regions, and how it evolves beyond its limitations as a place in which "the earth is very large and … we who dwell between the pillars of Hercules and the river Phasis live in a small part of it about the sea, like ants or frogs about a pond, and … many other people live in many other such regions" (Plato and Rowe, 1993). He continues:

> *For I believe there are in all directions on the earth many hollows of very various forms and sizes, into which the water and mist and air have run together; but the earth itself is pure and is situated in the pure heaven in which the stars are, the heaven which those who discourse about such matters call the ether; the water, mist and air are the sediment of this and flow together into the hollows of the earth.*

Plato is not the only one for whom ambiguity creates utopian potential. In his perception he does not identify anything particular to the region, rather, he argues, it is formed just like others elsewhere in the world, equally vaporous. His conception of these worlds within the world and their combination of critical relationships between history, geography, and identity allow new readings of historic frontiers in the Mediterranean. The true meaning of the contemporary Mediterranean, (especially) through the lens of historic readings, however, remains ambiguous. The Mediterranean in terms of its own history, I would argue, is an epistemological provocation in itself. It seems as if what holds it together also produces the divide; it is the place where the European Union is being tested to its breaking point, capitalism is being questioned, and where issues of migration, immigration, cultural and social inequality become most evident. It is also the region of (global) conflicts: the Israel-Palestine dispute, the wars in the Balkans, Libya, Egypt, Turkey, and the current civil war in Syria. Perhaps the reality of its own boundedness and the resulting crisis is what keeps the Mediterranean as a transitory space in constant tension. Italian philosopher Giorgio Agamben uses the analogy of an ill patient going to the doctor to describe the problematique:

> *'Crisis' in ancient medicine meant a judgment, when the doctor noted at the decisive moment whether the sick person would survive or die. The present understanding of crisis, on the other hand, refers to an enduring state. So this uncertainty is extended into the future, indefinitely. It is exactly the same with the theological sense; the Last Judgment was inseparable from the end of time. Today, however, judgment is divorced from the idea of resolution and repeatedly postponed. So the prospect of a decision is ever less, and an endless process of decision never concludes. (Agamben, 2013)*

The notion of crisis, and the decisive moment to determine whether the patient is sick, chronically ill, or going to die, is instrumental in the construction of the Mediterranean. Yet the way this illness, or as a matter of fact death, has been used to serve and legitimize political actions deprives the Mediterranean of a fixed identity, leaving citizens across the region without a clear understanding of who they actually are. Ironically, it seems as if what constituted the Roman *Mare Nostrum* still represents the axis around which the Mediterranean world revolves. But this time it is not ancient, and it distributes strife and controversy instead of a common rhythm (Vanstiphout, 2013). Reforms, continuous conflicts, and destabilizing demonstrations have shaken up the territory and, along with these, the emergence of new democracies with the support of new communication tools and social media endow the region with a continuous restructuring and transformation. These perpetual crises require continual deferment of judgment and of the resolution of identities it would allow.

Fernand Braudel has captured this nature of conflict best in his *longue durée*;[3] but conflict was also what he identified as keeping the region together. He employed the term "complementary enemies" to refer to powers condemned to living together and sharing the Mediterranean Sea, with wars and battles as the *courte durée* incidents in centuries-long periods of cohabitation (Braudel, 1995).

In the foreword to the English edition of 1972, Braudel vehemently repeated his claim against the analysis à la Pirenne in which the rise of Islam fatally broke the Mediterranean:

> *I retain the firm conviction that the Turkish Mediterranean lived and breathed with the same rhythms as the Christian, that the whole sea shared a common destiny, a heavy one indeed, with identical problems and general trends if not identical consequences. (Braudel, 1995)*

Political activist Predrag Matvejevic captures Braudel's claims and points out that boundaries in the Mediterranean, and therefore alternative conceptions, are drawn neither in space nor in time. He even goes so far as to assert:

> *There is in fact no way of drawing them: they are neither ethnic nor historical, state nor national; they are like a chalk circle that is constantly traced and erased, that the winds and waves, that obligations or inspirations, expand or reduce. (Matvejevic and Heim, 1999)*

While the Mediterranean's true definition as a system of interrelating systems is hard to decipher, the speeds at which new shifts and resulting spatial consequences occur have postulated a new level of significance for the readability of the region. Awareness lags behind the emergence of indefinable, shapeless regions devoid of identity, which underscores how current region-making processes in the Mediterranean, and elsewhere, are becoming increasingly transitory (Thierstein and Agnes 2008). Recent political conflicts have not only spawned new urban and regional morphologies; they also underline how the Mediterranean as a space is contributing to new readings of what is at stake in regionalism and urbanism on a

---

3 Braudel identifies three levels of time: the "longue durée," the "time" of geography; the "moyenne durée," the time of social, economic patterns and movements; and the "courte durée," the time of individuals and events.

much larger scale. The question is: is the Mediterranean of the twenty-first century being radically transformed by the very means that make it?

Whether through the lens of Horden and Purcell's kaleidoscopic liquidity or Braudel's stretching and layering of boundaries, the unfolding of history, geography, and identity is an interface of many systems and networks. The structural differences between these two methodological frameworks allow us to recover new proximities and qualities of a Mediterranean that is separated by politics and crisis. Problems or opportunities that existed within territorial and political boundaries of cities, regions, or even nation-states have radically shifting beyond borders. Some even argue that the limitations of borders in the Mediterranean are exaggerated. I propose that a possible perspective for looking at a new unfolding Mediterranean is by looking at the future of borders as a locus of innovation, especially in the context of environmental problems, energy, demographics, and infrastructural problems.

The complex history and the multifarious understandings and differentiations of its limitations make it difficult not to look at the region as a space with a "hundred frontiers" (Braudel, 1995). The spatio-temporal coordinates of points and lines that separate one place from the other are not simply geographically determined; nor are they understood as cultural constructions that belong to a "semantic field of similar formations across Africa, Asia, and Europe" (Chambers, 1995). It is not the hegemonic generalization, or the power over land and water that determines this large field of connected, interconnected, and unconnected urban, regional, and interregional structures. Rather, it is the conflict of "complementary enemies" (in Braudel's terms) that are condemned to live together and share the critical space. Cultural conflicts and economic competition have always been fundamental parts of coexistence in the Mediterranean. But historically, conflicts caused by invading powers always resulted in assimilation. Interestingly, this consisted not of assimilation to the dominant party, but assimilation within a deeper geographic logic that kept borders mobile and uncertain, and more "a horizon than a cartographic projection" (Braudel, 1995).

The distinctiveness and the meaning of the region, whether considered globally or regionally, are no longer defined by its natural frontiers (coast, islands, mountain ranges, the hinterland) that function as interpreters of signs and symptoms of historic and cultural values. Nor, if we refer to it as a lifestyle or ecology, are they a moral geography or a physical one; neither north nor south, Europe, Africa, nor Asia determine what the Mediterranean actually is. The liquidity of the frontiers in this uncentered historical *Kulturraum* are what offer new understandings of active limitations and a new sense of scalar dimensions within different territories all over the world. The complexity of the issues and conflicts, climate and the environment, and other new parameters makes it difficult to clearly differentiate the scalar frames that are based on cultural and economic constructs operating in a globalized world. Within these epistemological frames, the study of cities, regions, and worlds (within the world) provides new morphologies of what a newly unfolding Mediterranean can be today.

Urban and regional theories suggest that the constantly emerging, transnational urban and regional networks are the new control points in the geographically transforming, international "global city" in which the economy determines the locations and the aesthetics of power. The Mediterranean proves them wrong. The representation of the Mediterranean as de-territorialized does not reflect the region's unbounded complexity and diverse cultural ecology; nor can it indicate the endless debates about how to define it, or how its liquid geographies are bound together by the problems that constitute it. Whether we use frameworks such as ecology, migration, economy, or energy, the Mediterranean as such challenges every bounded conception of space. We also have to understand that notions of a limited set of relations—boundaries, cultures, etc.—are a paradox. How can a space or its definition be boundless, but yet be defined by all these limits?

GEOGRAPHY

In the early 1930s, an exhibition in Germany and Switzerland showcased one of the boldest architectural visions ever seen. Herman Sörgel, an architect and engineer from Munich, presented a totalizing vision on such a grand scale that it even outshined the fantasies of a novelist like Jules Verne. "Atlantropa," which was first known as "Panropa," was a project that re-imagined the entire Mediterranean region, from the Western Mediterranean and the Straits of Gibraltar to Israel in the East, and from Northern Italy to the Sahara desert in Africa.

A 35-kilometer-long dam in the Straits of Gibraltar was the key element in Sörgel's architectural utopia. The dams of Atlantropa disconnected the water supply from the Atlantic Ocean to the Mediterranean Sea and created a new super-geography of what was once known as the Mediterranean. Sörgel's vision of a gigantic dam gradually dried up the Mediterranean Sea and reduced the water level by 200 or more meters. The cultural geography affected by this shrinkage was the largest geographical transformation ever imagined. Sörgel's utopia created up to 600,000 square kilometers of new arable land where the sea had been. At the core of his design were several trans-Mediterranean arteries that supported the flow of people, cars, trains, and natural resources between civilizations inhabiting the Eurasian peninsula west of the Ural Mountains and Persia, including the Arabian Peninsula and Africa, beyond any historically predetermined infrastructural arteries.

Sörgel reimagined the Mediterranean by eliminating it. His project effaced the region's complex historic and cultural values; he re-designed every corner of the new region as a huge power plant with gigantic dam projects to provide an unlimited supply of energy. However, his *Gesamtkunstwerk* did not design the citizens for Atlantropa. Who would be the citizens populating these new territories?

His work was stimulated by the arts, but he was also a student of geography, which he believed was integral to architecture and his worldview. The German geographer Friedrich Ratzel and the philosopher Oswald Spengler were elemental to Sörgel's self-assessment as *Weltbauer* [global architect]. Both shaped this understanding in very different ways. Ratzel was fundamental to Sörgel's concept

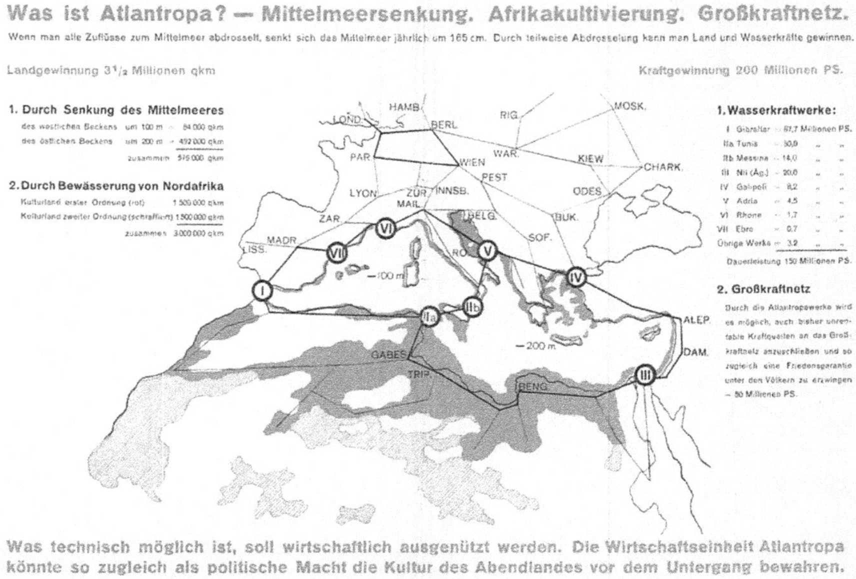

11.2 What is Atlantropa? (1931). Herman Sörgel: Atlantropa. Zürich, München 1932, p. 38. Reproduced courtesy of Deutsches Museum.

of an architect as possessing a chiliastic spirit that liberated him to aim for cosmic depth, and a pacifist spirit that aspired to create new borders. Ratzel's theory was that of *der Staat als Organismus* (the state as organism), in which he equated dams, autobahns, railways, and bridges to the digestive and circulatory systems in natural organisms. Ratzel argued that the more extensively and qualitatively these systems were built, the more the organism, the *Lebensraum* (living space) of a nation-state, would thrive.

Spengler, on the other hand, had a much deeper and more personal influence on Sörgel. Both were friends from the time they lived in Schwabing. In 1918, Spengler published his book, *Der Untergang des Abendlandes* [*The Decline of the West*], in which his pessimistic prophecies about the extinction of western culture through "civilization" and "overpopulation" captured Sörgel's attention (Spengler, 1980). Sörgel, however, had a more positive view and believed in the technical advancements of modern societies. He shared Spengler's critique of the nineteenth century and that the uncontrollably growing urban environment was a sign of the weakness of western civilizations. But in contrast, he believed in technical and artistic urban planning solutions for the future. While Spengler looked at technology as a demonic force that destroyed culture, Sörgel saw in technology the key to its salvation. Along with Bruno Taut and Le Corbusier, Sörgel actively promoted *Weltbauen* in large-scale dimensions, striving for global solutions that questioned geography, including human activity, existing distribution patters, population, and resources, and replaced it with utopian imaginations. The use of technology to address the challenges of future developments and achieve unprecedented improvements through the alteration of global geographies characterized their

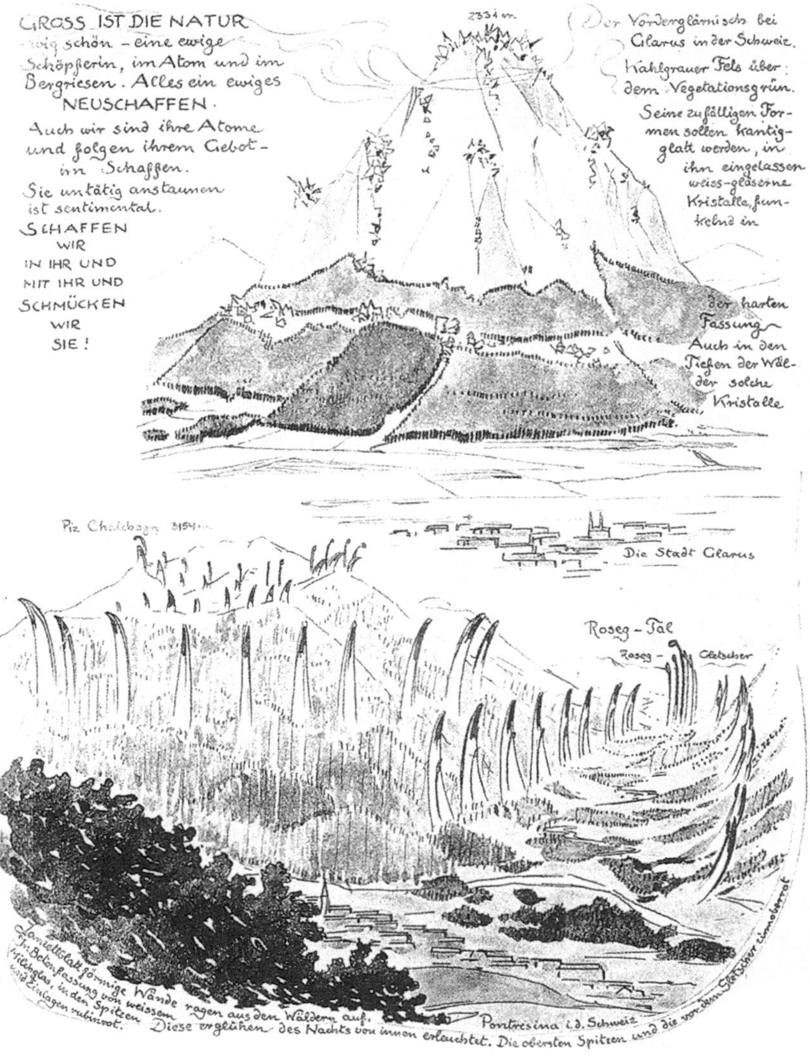

11.3 Weltbauen, Bruno Taut, "Alpine Architecture" (1918).

ideas as *Weltbaumeister*. All of these activities looked at planning as a totality that extended to every corner of the planet. In one of his countless publications and movies, *Atlantropa: Der neue Erdteil, Das Land der Zukunft* [*Atlantropa: The New Continent, Land of the Future*], Sörgel responded to the Spenglerian despair of civilization with his engineering megalomania.

Atlantropa's objective was not only to become a new supercontinent centered in the Mediterranean, formed from Europe and Africa, but also to solve all major problems of a continent battered by crisis after crisis. With the Great Depression, World War I, and a looming World War II, Europe was in need of a vision for the future. However, Sörgel's angle on the problem was purely technocratic. He was convinced that in order to remain globally competitive with technologically and economically advanced America and emerging Pan-Asian territories, Europe needed to be self-sufficient which, according to Sörgel, required possession of territories in all climate zones. To position Europe as a sustainable global player among the United States of America, Great Britain—which he believed could not maintain its empire in the long run—and Asia—a mystery to Europeans—his plan was to lower the Mediterranean basin and produce infinite amounts of electric power through his dam. As a result, vast tracts of land would have been generated for new urban settlements and agriculture, including the Sahara desert, which was to be irrigated with the help of three sea-sized, man-made lakes throughout Africa. "The massive public works, envisioned to go on for more than a hundred years, would have relieved the unemployment crisis on the continent and the acquisition of new land would ease the pressure of overpopulation, which Sörgel thought was the fundamental cause of political unrest in Europe" (Group Edit Suisse, 2003). The polemic of his project was not only to change the geography of the entire Mediterranean, but also of the African continent. The post World War I racist ideologies, which saw Africa as an empty continent devoid of history and culture, were integrated in Sörgel's belief in which he saw technology as a political power (Voigt, 2007). His vision was an alternative to anything ever imagined, and with the larger geopolitical goal of venturing deep into the Congo Basin he opened Pandora's box by wanting to secure the vast natural resources of Africa. His idea was to exploit Africa with European technological and engineering know-how and turn the continent into a "territory actually useful to Europe" (Voigt, 2003, p. 29).

At a time in which apocalyptic visions determined the Zeitgeist and the politics of imperialism were dominant, he placed his project within a larger geopolitical order. He followed a Social Darwinist and colonialist school of thought, declaring, "The fight for survival is a fight for territory" (Voigt, 2007, p. 29). His plans for Atlantropa not only aimed to revolutionize the north-south connection between Europe and Africa, but he also aimed to transfigure the global west–east imbalance into a "harmonious coexistence" of the three A's: America, Asia, and Atlantropa, which he considered *Kontinentale Grossräume* (continental megaspaces) that would coexist in a new world order of supercontinents. As a result, Atlantropa not only completely obliterated what was known as the Mediterranean, but it completely sacrificed cultural complexity for the promise of a continental European economic security and energy independence by regulating the flow between the Atlantic Ocean and the Mediterranean Sea and its cultures. Inadvertently, Atlantropa was advocated as a site of a new world order of geographic, geopolitical, and architectural visions.

In its early stages, Atlantropa was created in the political context of the Pan-European Union, founded in 1923. The assembly brought together by the Austrian (geo)-politician Richard Coudenhove-Kalergi sought political unification of the European continent after the devastating First World War. When in 1929, Sörgel's vision took on the name Panropa, he wanted to emphasize its close connection to the Pan-European Union. However, in 1932 Sörgel replaced Panropa with the official project name, Atlantropa, to avoid confusion with the Pan-European Union. He changed the name primarily because he did not believe in the ideological principles of the Pan-European Union, and disagreed with Coudenhove-Kalergi politically and morally. Sörgel believed that only an economic union could bring Europe together, and advertised the guaranteed profit of Atlantropa and its economic benefits and energy independence. The change of name was also an expression of the ways in which Sörgel's geopolitical ambitions went far beyond Coudenhove-Kalergi's organization. Atlantropa, the term invented by Sörgel, meant new territory at the Atlantic Ocean [*Festland am Atlantik*], and stood for the idea of politically and geographically uniting Europe with Africa into a supercontinent of tomorrow.

IDENTITY

Many decades after Atlantropa, the Dutch avant-garde architect Rem Koolhaas and his firm *Office for Metropolitan Architecture* (OMA)—alongside AMO, the firm's research-based think tank within OMA—proposed another "megalomaniac" vision for Europa. Koolhaas and Reinier De Graaf, his partner in charge of the project, envisioned "Roadmap 2050," a redesign of the entire European continent, including North Africa, along new energy-saving frontiers. The project promises a prosperous "United Nations of Energy" (Hartman, 2010) in a new energy-independent and decarbonized European continent. Typically for OMA/AMO, "Megalomania" says de Graaf, is a standard part of the repertoire of the firm.

The core of their proposal, however, is "Eneropa," a new map for Europe that completely restructured the continent under one idea of renewable energy by redrawing national boundaries and replacing nation-states with regions based on energy sources supplying the larger Eneropa grid. The OMA/AMO study reorganized the entire continent as if it were a blank slate and developed a new energy-based identity for Europe: Eneropa, unlike the European Union (EU), is not defined by economic or cultural identities. Rather, similar to Sörgel's Atlantropa, the proposal overlooks the complex cultural identities of Western and Eastern Europe, and instead develops an alternative scenario for the entire continent. Eneropa reorganizes Europe into regions defined by their available energy source, which, ideally, contributes to a larger energy grid. Mediterranean nations such as Portugal, Spain, Italy, and Greece are combined into the energy territory *Solaria*; Central European states such as Germany are *Geothermalia*, and others such as Switzerland, parts of France, and Austria are *Enhanced Geothermalia*; parts of Great Britain and Ireland are *Tidal States*; the Scandinavian nations are *Isles of Wind* in the North Sea; the project even reunites Former Yugoslavia into *Biomassburg*.

The idea is simple: Eneropa is a redesign of the entire European territory based on the creation of a power network linking abundant, regional, renewable energy sources that contribute to and compensate for each other within a large renewable energy network. For example, wind is abundant in Northern Europe, sun in the Mediterranean, waterpower in the Alps, and in different seasons and different quantities. However, if it were windless in the North but sunny in the Mediterranean, the Eneropa power network would compensate and balance

11.4 Eneropa. © OMA.

deficiencies in areas where energy was needed. The 2050 report also discusses the possibility of expanding Eneropa into North Africa in an energy exchange utilizing the region's solar potential in return for wind energy from Eneropa's Isles of Wind.

The big idea of Eneropa, however, is a graphic narrative. The firm conceptualized and visualized geographic, political, and cultural implications of the integrated and decarbonized European power network into a proposal that operates in areas far beyond the traditional boundaries of architecture. Koolhaas and OMA/AMO are known for pushing these boundaries. His global engagement in research, development, preservation, and politics manifests his broad inquiry across disciplines. For a few years, Rem Koolhaas worked as a member of the EU's Reflection Group on similar schemes. Eneropa is just a continuation of his larger interests, and two projects AMO has been working on for several years stand out as foundational.

In 2004, the president of the European Commission, Romano Prodi, and Belgian Prime Minister Guy Verhofstadt invited AMO to develop proposals to best meet the needs, functions, and identity problems of a European capital. AMO's proposal addressed the presence of the EU through the architecture of its institutions, but also saw an opportunity to address the representation of Europe at large; other than the European flag, there was no visual narrative that connected the continent.

In their proposal, AMO pointed out Europe's iconographic deficit. The visual form of the EU's flag was a result of what they interpreted as "a widespread ignorance about the activities and the origins of the EU among the general public" (OMA, 2001). In a series of experiments, the AMO think-tank came up with illustrations and "image bites" in which all the flags of the EU member states were merged into a single pattern. The result resembled a colorful barcode of European nations that only represent a sliver of each individual cultural identity while suggesting the functional advantages of acting together. The critique of the old flag was that the number of stars representing member states was fixed. OMA's barcode had the advantage of being expandable infinitely as new members join the economic union. However, the barcode also signifies an optically- readable machine representation of data in relation to an object, and not a cultural entity, which in the proposed context seems ironic.

Another pre-Eneropa project was a 2008 masterplan for offshore wind farms in the North Sea. The main concept here was based on the urgency of calls for sustainable and secure energy requiring a collective mobilization of intelligence and ambition that would exceed standard piecemeal solutions to climate change. The project mapped out a large-scale renewable energy infrastructure in the sea for supply, distribution, and strategic growth, engaging with all its surrounding countries, and potentially those beyond, in "a supra-national effort that will be both immediately exploitable and conducive to decades of coordinated development" (OMA, 2008). The masterplan also included infrastructure to support manufacturing and research, as well as existing marine life. Koolhaas argued that all North Sea countries were uniquely positioned to benefit from large-scale wind farms not only in regard to the economic benefits of renewable energy production, but also from research and development in the advancement of wind, wave, tidal, and biomass energy production.

Ultimately, Koolhaas' ambitions come together in his interest in policy-making. His proposal not only addresses new cultural geographies of the European continent, and potentially Africa, but it also is a policy roadmap to reduce CO_2 emissions by 80–95 percent by 2050. Together with his partners McKinsey, Imperial College London, the energy consultancy KEMA, and analyst Oxford Economics, AMO mapped out scenarios for meeting this target. The technical and economic analyses outlines why a zero-carbon power sector is required to meet the commitment of reduced CO_2 emissions and, given current standards of technology, illustrates its feasibility by 2050. The goal can be achieved through the complete integration and synchronization of the EU's energy infrastructure, and the plan shows how Europe can take maximum advantage of its geographical diversity to ensure energy security for future generations.

Additionally, in this typically provocative, OMA'esque sense, the report raises the question of what to do with the old European energy infrastructure. Over the past 20 years, European nations have invested billions of dollars to revive relics from the industrial age and bring them into the twenty-first century. The report suggests that the old infrastructure could be preserved as UNESCO sites of a pre-decarbonized Eneropa, and function as a cultural memorial for the larger world. Thus, the old European identity would be transformed into a new, post-carbon narrative.

There is also the question of how much this would cost. The study has produced figures that show that the scheme would not cost much per capita, especially when compared with road building, war in Iraq, or bailing out bankers. It also points out the benefits of reducing reliance on nuclear power, Middle Eastern oil, and Russian gas. The overall conclusion is that the economic benefits would outweigh the cost, which could be reduced by 80 percent, especially if North Africa and its abundance of sun could be integrated into the Eneropa grid.

ALTERNATIVE CONCEPTIONS

Herman Sörgel's and Rem Koolhaas' super-geographies envisioned territories so alternative to the cultural landscape of the Mediterranean, and so divergent from scholars' perceptions, that they fundamentally transformed what ultimately determined it. Their large-scale frameworks and the imposing of new territorial structures radically eliminated "the Mediterranean" and replaced its regional, cultural, and geographic fabric with the promise of new world orders.

Undoubtedly, in understanding how the region is constructed Braudel, Horden and Purcell provided insight into the history and geography of the Mediterranean, and how both conflict and the emergence of limits materialize from concrete networks of causes and effects in relation to a larger identity. Furthermore, their scholarship identifies the presence of frontiers as much by what crosses them or what holds them apart. But Sörgel's and Koolhaas' proposals both suspend these traditional perceptions, understandings and conceptions of borders and frontiers that are based on geography and culture. Instead, they propose totalizing visions in which the solution for larger geopolitical questions, as well as the ones of culture and identity, were sought in questioning geography and replacing it with utopian imaginations.

In the last 20 years, new developments, from the reforms of the Arab Spring's destabilizing demonstrations to the deliberative discussions about a Mediterranean Union, have further altered the territory, and along with the emergence of new democracies, endowed the region with unremitting reorganization, transition, and transformation. From Egypt's Tahrir Square to Athens' Syntagma Square, and from Madrid's Puerta del Sol to Algiers' Habib Bourguiba Avenue, mainly urban revolutions have recast the image of the Mediterranean and presented it with a new significance and a new sense for urban and transnational identities. The revolts against regimes, global economics, and capitalism not only turned the dissolving European-dominated identity of the Mediterranean into a volatile and multifarious "Medi-terranean," but they also reaffirmed that the territory and its frontiers are highly contested.

Points and lines—bound, unbound, connected, or interconnected, and as Giorgio Agamben argues, with every culture's identity sitting on both sides of its frontiers—separating one place from another are not simply geographically determined; neither are they dependent on utopian imaginations or interventions privileging one Mediterranean identity over another. Current events actually show how tedious transformative processes are. Political unrest and the shifting of social, cultural, political, economic and ultimately environmental borders are a reflection of the fluidity of the Mediterranean territory today. These events are a testament to real transformations, rather than Sörgel's and Koolhaas' metanarratives. However, there is something to be said about the radical nature of their proposals. Both never question the larger territorial coherence of the Mediterranean (understanding that there is no such thing in the Mediterranean). Quite on the contrary, they understand the region as a geography that is neither determined as Mediterranean nor Medi-terranean, maintaining certain regimes of demographic control along political and cultural frontiers.

The true Mediterranean unfolds somewhere between the modern frameworks set by Sörgel and Koolhaas and the continuous political unrest. Drawings and sketches by Peter Fend[4] or Philippe Rekacewicz, or Max Ernst' painting *Europe after the Rain I*, or contemporary photography by Bas Princen[5] and Iwan Baan present the territorial, and I argue, cultural fluidity of the Mediterranean in ways that render borders and boundaries less as objects than as specific contextual geographies with hybrid meanings. The works of these artists amorphously reveal (altered)

---

4 Maps in: Petrov, A. 2013. *The Mediterranean. New Geographies Vol. 5*. Cambridge, MA: Harvard University Press, pp. 126–8.

5 Princen, B. "A Mediterranean Photodossier." In Petrov, A. 2013. *New Geographies 5 "the Mediterranean"*. Cambridge, MA: Harvard University Press, pp. 179–94.

11.5 Max Ernst's painting "Europe after the Rain I" (1933). © 2014 Artists Rights Society (ARS), New York/ADAGP, Paris.

territories that utilize ambiguity as a potential means of negotiating place not only between notions of centrality and peripherality, but also social and cultural identities on one or the other side of the frontier. The metaphorical power of points, lines, pixels, and rain (in Ernst's painting[6]) offer enough *Spielraum* (space/leverage) to construct a new synthesis of an unbounded *Kulturraum* in which understandings of active limitations and a new sense of scalar dimensions emerge within the larger Mediterranean territory. In this sense, Predrag Matvejević's idea of borders in which the chalk circles are constantly traced and erased gain a new importance, and Ernst's metaphoric rain becomes the mediator that assimilates the entire territory, not just its politics.

The argument for a conceptual ambiguity, as seen not only through the eyes of the artists or planners but also Mediterraneans, and the real pertinence of the Mediterranean is not that of a geopolitical zone, or of a grouping of nation-states with clear borders and a sea around which everything is organized. The artwork's conceptual ambiguity argues for an unbounded Mediterraneanity, seen through the eyes of projected desires and hybrid interpretations, with Mediterranean

[6] The title dates back to an earlier painting sculpted from plaster and oil (and painted on plywood from the set of L'Âge d'or) to create an imaginary relief map of a remodeled Europe completed in 1933, the year Hitler took power. In *Europe after the Rain I*, Max Ernst's depicts an emotional desolation, physical exhaustion, and the fears of the destructive power of total warfare combined—after the rain of fire, the biblical deluge, and the reign of terror.

frontiers that are not just sites of new "world" orders, but also pivot points between geographical and utopian imaginations. The geographic limits of the Mediterranean, whether determined by history, historians, or political unrest, will always be "Mediterranean," whether they are eliminated or not. The Mediterranean will always continue to be an expression of both the "geographic" and the distinct individuals that are instrumental in its own construction.

REFERENCES

Abulafia, D. 2003. *The Mediterranean in History*. Los Angeles, CA: J. Paul Getty Museum.

Agamben, G. 2013. "The Endless Crisis as an Instrument of Power: In conversation with Giorgio Agamben" [online]. Available at: http://www.versobooks.com/blogs/1318-the-endless-crisis-as-an-instrument-of-power-in-conversation-with-giorgio-agamben (accessed 4 June 2013).

Braudel, F. 1995. *The Mediterranean and the Mediterranean World in the Age of Philip II*. Berkeley, CA: University of California Press.

Chambers, I. 2013. "Space, Memory, Rhythm, and Time: Constructing a Mediterranean Archive." In Petrov, A. (ed.) 2013. *New Geographies*. Cambridge, MA: Harvard University Press.

Figure 11.4. Office for Metropolitan Architecture. Heer Bokelweg 149, 3032 AD Rotterdam, Netherlands.

Group, E.S. 2003. *Atlantropa*. Cabinet.

Hartman, H. 2010. "Welcome to Eneropa, Rem Koolhaas' vision of a united nations of energy." *The Architectural Review*.

Horden, P. and Purcell, N. 2000. *The Corrupting Sea: A Study of Mediterranean History*. Oxford, UK/Malden, MA: Blackwell.

Lahoud, A. 2013. "The Mediterranean: A New Imaginary." In Petrov, A. (ed.) 2013. *New Geographies 5 "the Mediterranean"*. Cambridge, MA: Harvard University Press.

Matvejevic, P. and Heim, M.H. 1999. *Mediterranean: A Cultural Landscape*. Berkeley, CA/London: University of California Press.

OMA. 2001. http://www.oma.eu/projects/2001/eu-barcode.

OMA. 2008. http://www.oma.eu/projects/2008/zeekracht.

Petrov, A. 2013. *The Mediterranean: Worlds, Regions, Cities, and Architectures*. Cambridge, MA: Harvard University Press.

Plato and Rowe, C.J. 1993. *Phaedo*. Cambridge, UK/New York, NY: Cambridge University Press.

Purcell, N. 2003. "The Boundless Sea of Unlikeness? On Defining the Mediterranean." *Mediterranean Historical Review*, 18, 9–29.

Sarkis, H., Herzfeld, M., and Ben-Yehoyada, N. 2013. "Sea of Scales and Segments." In Petrov, A. (ed.) 2013. *New Geographies 5 "the Mediterranean"*. Cambridge, MA: Harvard University Press.

Spengler, O. 1980. *Der Untergang des Abendlandes: Umrisse einer Morphologie der Weltgeschichte*. Mèunchen: C.H. Beck.

Thierstein, A. and Fèorster, A. 2008. *The Image and the Region—Making Mega-City Regions Visible!* Baden: Lars Muller Publishers.

Vanstiphout, W. 2013. "Braudel's Donkey: Historians and the Mediterranean as a Political Project." In Petrov, A. (ed.) 2013. *New Geographies 5 "the Mediterranean"*. Cambridge, MA: Harvard University Press.

Voigt, W. 2007. *Atlantropa: Weltbauten am Meer. Ein Architekturtraum der Moderne, Hamburg, Dölling and Gallitz*. M.A.T. Music Theme and Licensing.

12

The NSK State and the Collective Imaginary

Conor McGrady

INTRODUCTION

As a project of the NSK (Neue Slowenische Kunst) artists' collective, the NSK State in Time emerged shortly after the collapse of socialism in Eastern Europe. This collective grew out of the social and cultural conditions of Yugoslavia in the 1980s. In the wake of long-time leader Marshall Josip Broz Tito's death, Slovene youth and underground culture clashed with the Yugoslav authorities, while power struggles and rising nationalism across Yugoslavia began to threaten the federation. NSK co-founders and multi-media art/industrial music group Laibach faced censorship for their visual and aural reprocessing of art, ideology and politics. As events in the 1980s accelerated towards the disintegration of Yugoslavia in war, NSK emerged as a multi-faceted entity, with Laibach's controversial and provocative presence now framed within an expanded collective structure consisting of the artists collective IRWIN, Scipion Nasice Sisters Theatre group, New Collectivism design and a Department of Pure and Applied Philosophy. The collective's characteristic appropriation of authoritarianism merged nationalist and totalitarian iconography with pop art and the aesthetics of the early twentieth century avant-gardes. NSK virtually seceded from newly independent Slovenia in 1992 to become a state in time and without borders—a utopian social sculpture embodying a symbolic transcendence of the nationalism engulfing the region. Shortly after its founding, the state began to issue NSK passports and open temporary embassies in apartments and galleries in numerous locations, including Moscow, Sarajevo, Berlin, Ghent, Glasgow and Dublin.

ART AND IDEOLOGY AND PROVOCATION

The NSK State takes the concept of the state as a Duchampian ready-made and merges it with Joseph Beuys' ideas of social sculpture. In reprocessing early twentieth century utopian idealism as one of multiple elements within its framework, it represents an ideal form of state—a de-territorialized social organism that provides a conceptual alternative to the politics of alienation generated by contemporary political systems and modern nation states. But unlike other alternative social formations, the signifiers of authoritarianism that have characterized the work of NSK and its constituent groups since the formation of the collective in 1984 undercut its utopian principles. In this sense, the NSK State is not only an alternative virtual social formation, but a critical lens through which to view the relationship between art and ideology. The tension between autonomy and hegemony as played out in the foundation of alternative social, political and cultural spaces is implicit in its actions and symbolic gestures. As an art project that temporarily interpolates itself into existing social and political situations it operates as a methodological field in which social relations are explored, tested and challenged. The complexity of place and belonging that is integral to the formation, disintegration and re-formation of modern nation states comes under scrutiny in the manifestations of the NSK State and the actions and initiatives of its citizens, who number over 15,000.

THE BODY INHABITED BY THE STATE

Just as the NSK State appropriates the utopian desires of early modernism, it does likewise with the totalizing narratives of nationalism and repressed authoritarian violence that lay at the heart of the modern nation state. For example, the symbolic appropriation of multi-national armed forces, as carried out in IRWIN's NSK Garda actions, evokes the controlled and ritualized threat of violence that characterizes the modern nation state. Rows of rigid, uniformed bodies in military formation echo the physical walls and barriers that represent architectural manifestations of statehood, including those that enclose and exclude in contemporary zones of contention. The defensive bulwark formed by the mechanized body prevents the dissolution of the self in the morass, characterized by Klaus Theweleit (1989) as a miasmic sea of others that constitutes the enemy within as well as without. The uniformed body, to paraphrase Allen Feldman (1991), is the body infested and inhabited by the state. The state is internalized to such a degree that its subject thinks and acts in accordance with its implicit and explicit commands. Woven into the virtual and physical manifestations of the NSK state, this language of force is indicative of the sociological chains tying us to abstract geopolitical retro-formations. If the NSK State represents the possibility of a return to the commons, its incorporation of the politics of enclosure imbues it with a pervasive sense of ambiguity and tension.

The engagement with history, politics and ideology in NSK State aesthetics and actions provides a perpetual pull between alternating poles of utopia and dystopia. Actions that appropriate the uniformed military personnel of other nation states in temporary allegiance to a virtual state, amplify the tension between fear and attraction that characterizes any relationship to authority. Interestingly, NSK citizen generated Folk Art—artwork generated by NSK Citizens that remixes NSK-related themes and iconography—for the most part eschews the utopian dynamic present in the NSK State in favor of the dystopian. The majority of such works tend to address the aesthetics of power, exposing the latent authoritarianism at the heart of every political system.

The issue of power poses the perennial conundrum involved in building any alternative society, from the Temporary Autonomous Zone to the commune, squat or more recently, micronation. Cold war polarization, third world liberation movements and the waves of social upheaval that swept through the late 60s rekindled the desire for alternative forms of social organization. Alongside the re-imagined communities generated by the communal and squatter movements, Micronations have spread rapidly in recent years. As independent nations or states that are not recognized by world governments or major international bodies, they usually exist as social or political simulations. At the first NSK Citizens' Congress in Berlin in 2010, it was discussed as to whether or not the NSK State should formally align with the micronation movement. Delegates at the Congress argued that the State in Time transcends micronations, as they tend to mimic reactionary forms of government, including fiefdoms, monarchies and other feudal structures. The establishment of bilateral relations with micronations was ultimately deemed to be in contradiction to the declamatory status of the NSK State as "the first global state of the universe?"

BEYOND THE "CAGE OF NATIONAL CULTURE"

The subversive power of the NSK state lays in its lack of fixed points of reference—its absolutism yet its open-endedness. It is within its abstract conceptualization of space that citizens have the ability to belong to a de-territorialized social organism. As an artistic project it questions how we relate to the collective imaginaries of nation states and territories in an era characterized by globalization, neo-liberalism and crisis. Symbolically, its unlimited spatial dimension replaces concrete borders with a collective flux of ideas that moves beyond the constricting parameters of what founding NSK member Eda Cufer refers to as the "cage of national culture."

The actions of the NSK State resonate with the development of socially engaged artistic practice and the recent Occupy movement. The zones of occupation that sprung up around the Occupy movement reclaim the impetus of previous social movements to test new ways of living. While Occupy relied on the appropriation of physical space, the NSK state symbolically supersedes nations, boundaries and territories. Encompassing the totality of the work of its founders and the decentralized initiatives of its citizens, the NSK State constitutes a critical if ambiguous alternative. Its evocation of post-national social relations remains prescient in an age paradoxically characterized by the weakening power of the nation state under neoliberalism alongside the continued reassertion of nationalism and state control. The importance of the passport embodies this paradox, as a document of identification and belonging as well as exclusion. The NSK passport provides access to a hybrid and provocative zone that reflects this complexity. In symbolizing the conscious extraction of the individual from the unconscious imposition of ideology, citizenship of the NSK State tests the boundaries of the collective imaginary, while at the same time questioning the potential for decommissioning the police states that we carry around in our heads.

REFERENCES

Feldman, A. 1991. *Formations of Violence*. Chicago: University of Chicago.

Theweleit, K. 1989. *Male Fantasies Vol 2. Male Bodies: Psychoanalyzing the White Terror*. Cambridge: Polity.

12.1 IRWIN, NSK Embassy Moscow. Regina Gallery/Apt Art, Moscow. Metal, silkscreen, 82 × 52 × 3.5 cm, 1992. Photograph courtesy of IRWIN.

12.2 NSK State Sarajevo, National Theatre Sarajevo, 1995. Photograph courtesy of IRWIN/New Collectivism.

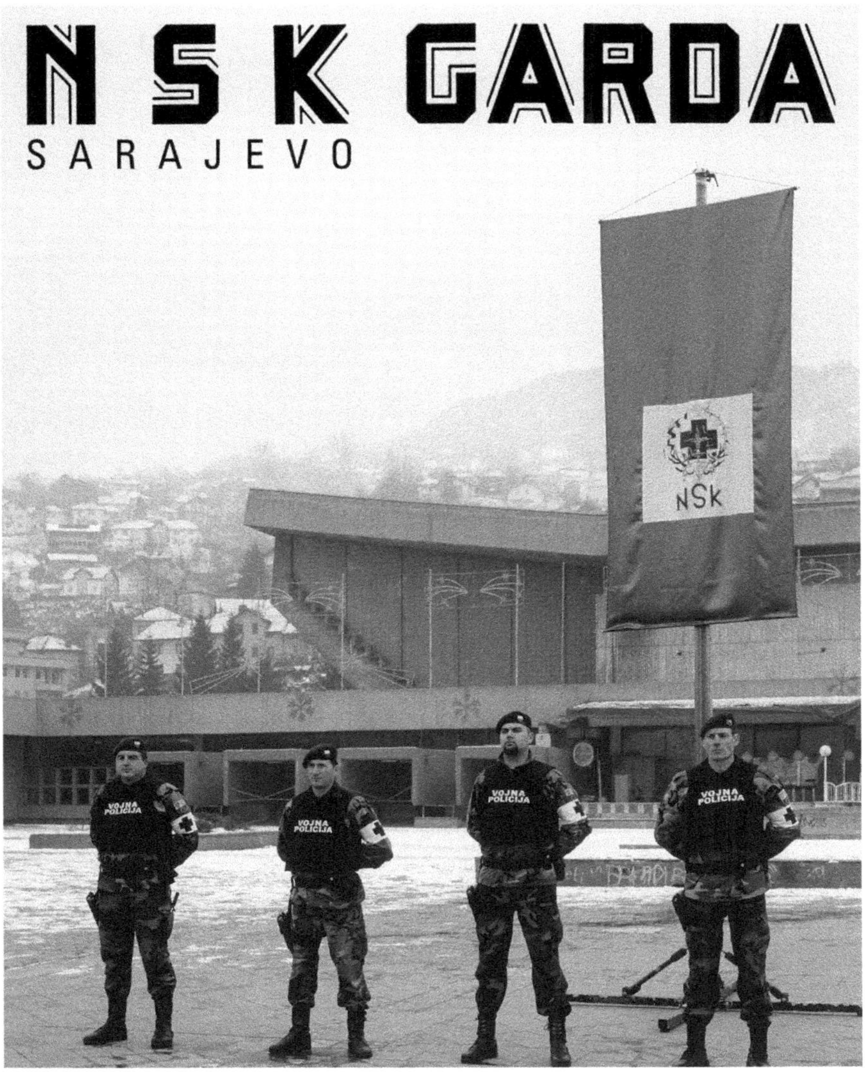

12.3 NSK Garda Sarajevo. IRWIN, in collaboration with Bosnian Army, 2006, Iris print. Photograph by Igor Andjelič. Courtesy of Galerija Gregor Podnar.

12.4 NSK Garda Sligo. IRWIN, in collaboration with Irish Army, 2010, Iris print. Photograph, IRWIN archive. Courtesy of Galerija Gregor Podnar.

12.5 Delegates, NSK members, Congress team members and guests of the "First NSK Citizens' Congress," Haus Der Kulteren Der Welt, Berlin, October, 2010. Photograph by Wig Worland. Courtesy of IRWIN.

12.6 Passport stamp for NSK Passport Office, MoMA, New York, 2012. Design by david K. Thompson. Image courtesy of the artist and NSK New York Organizing Committee.

12.7 david K. Thompson, Occupied Sates: Nova Akropola, 2011.
Image courtesy of the artist.

12.8 IRWIN, NSK Passport Office at MoMA, 2012. Photograph by Paula Court. Courtesy of the Museum of Modern Art.

12.9 NSK State Passport.
Photograph courtesy of IRWIN/New Collectivism.

12.10 IRWIN, Time for a New State, Lagos, Nigeria 2010. Photograph courtesy of IRWIN.

Index

For Product Safety Concerns and Information please contact our EU
representative GPSR@taylorandfrancis.com
Taylor & Francis Verlag GmbH, Kaufingerstraße 24, 80331 München, Germany

www.ingramcontent.com/pod-product-compliance
Ingram Content Group UK Ltd.
Pitfield, Milton Keynes, MK11 3LW, UK
UKHW050759180926
475701UK00008B/1137
* 9 7 8 0 3 6 7 7 8 7 3 4 9 *